Propagation and Flowering of Ornamental Plants

Propagation and Flowering of Ornamental Plants

Guest Editors

Yong Ha Rhie
Seung Youn Lee
Jongyun Kim

MDPI

Basel • Beijing • Wuhan • Barcelona • Belgrade • Novi Sad • Cluj • Manchester

Guest Editors

Yong Ha Rhie
Department of Horticulture
Kangwon National University
Chuncheon
Republic of Korea

Seung Youn Lee
Department of Smart Horticultural Science
Gyeongkuk National University
Andong
Republic of Korea

Jongyun Kim
Department of Plant Biotechnology
Korea University
Seoul
Republic of Korea

Editorial Office
MDPI AG
Grosspeteranlage 5
4052 Basel, Switzerland

This is a reprint of the Special Issue, published open access by the journal *Horticulturae* (ISSN 2311-7524), freely accessible at: https://www.mdpi.com/journal/horticulturae/special_issues/TR6IID1B17.

For citation purposes, cite each article independently as indicated on the article page online and as indicated below:

Lastname, A.A.; Lastname, B.B. Article Title. *Journal Name* **Year**, *Volume Number*, Page Range.

ISBN 978-3-7258-7212-1 (Hbk)
ISBN 978-3-7258-7213-8 (PDF)
https://doi.org/10.3390/books978-3-7258-7213-8

Contents

horticulturae

MDPI

Article

Metabolomic Analysis of *Lycoris radiata* across Developmental and Dormancy Stages

Xueru Jiang [1], Xuying Wei [2], Hua Cheng [1], Xin You [1] and Junhuo Cai [1,*]

1 Jiangxi Provincial Key Laboratory of Conservation Biology, Jiangxi Provincial Key Laboratory of Subtropical Forest Resources Cultivation, College of Forestry/College of Art and Landscape, Jiangxi Agricultural University, Nanchang 330045, China; hnfgjxr@163.com (X.J.); chenghua0802@sina.com (H.C.); youxin0428@163.com (X.Y.)

2 College of Art, Jiangxi Finance and Economics University, Nanchang 330032, China; weixuyin74@163.com

* Correspondence: cjhuo7692@163.com

Abstract: The *Lycoris radiata* (L' Herit.) Herb. is a perennial bulbous plant characterized by its high ornamental and medicinal value, exhibiting a unique growth rhythm where the flower and leaf do not coexist and a period of summer dormancy. However, its metabolic response to various developmental stages remains unclear. To address this gap, we conducted a non-targeted metabolomic analysis spanning six developmental stages of *L. radiata*. The results showed that most differentially accumulated metabolites (DAMs) demonstrated enrichment predominantly in carbohydrate and amino acid metabolism pathways, with the former being more active during vegetative growth and the latter during reproductive stages. The proportion of DAMs categorized under "quaternary ammonium salts", "tricarboxylic acids and derivatives", "fatty acids and conjugates", and "pyrimidine nucleotide sugars" was notably higher in comparisons between the flowering and dormancy stages than in other comparative groups. Furthermore, DAMs involved in the KEGG pathways of C5-branched dibasic acid metabolism and lysine biosynthesis were uniquely identified during the transition from Dormancy to Flowering. The proportion of DAMs associated with "linoleic acids and derivatives" and "pyridines and pyridine derivatives" was notably higher in the leafing out versus flowering comparison than in other comparative groups. Furthermore, the glycolysis/gluconeogenesis pathway was uniquely enriched by DAMs during this phase. This study provided an in-depth view of metabolite changes in *L. radiata* over its annual growth cycle, enriching our understanding of the regulatory mechanisms governing its development, dormancy, and flowering.

Keywords: *Lycoris radiata*; metabolomic; developmental; dormancy; flowering

1. Introduction

Lycoris radiata (L' Herit.) Herb, a perennial bulbous plant within the Amaryllidaceae family, predominantly thrives in East Asia's warm temperate and subtropical regions. This species is esteemed for its substantial ornamental and medicinal virtues. Alkaloids extracted from the bulbs of *Lycoris*, such as galanthamine, lycoramine, and lycorine, hold potential for pharmaceutical development, targeting an array of diseases [1–3]. Additionally, due to its distinctive and vibrant flowers, coupled with robust stress and shade tolerance, *L. radiata* frequently serves in landscaping as a ground cover in shaded gardens, as well as in containers and as fresh-cut flowers. Currently, the integration of large-scale cultivation with tourism and the production of medicinal substances yields considerable economic benefits. Nonetheless, the commercial scalability of *L. radiata* is constrained by challenges in effectively regulating its flowering period. Investigating its biological traits and flowering mechanisms is pivotal for advancing flowering regulation technologies.

L. radiata exhibits the unique biological characteristic of "flowers and leaves not meeting"; it enters dormancy in summer, blooms in autumn without leaves, and subsequently produces leaves that remain through autumn, winter, and spring [4]. Simplifying, the life cycle of *L. radiata* encompasses a leaf stage and a leafless stage. The leaf stage extends from October to May of the subsequent year and is subdivided into four periods: leaf emergence (October), rapid leaf growth (November), leaf maturity (December to February), and leaf senescence (March to May). The leafless stage comprises a dormancy period (June to July) and a flowering period (August to September) [5]. Flower development can be further categorized into the flower bud differentiation period (April to May), flower organ development period (June to July), scape formation (August), and flowering period (September) [6]. Flower bud differentiation occurs as the leaves begin to yellow, while flower organ development coincides with the dormancy period [6].

Recent research on the development and flowering of *Lycoris* primarily investigates the sink-source transition of carbohydrates within bulbs. Key areas of focus include the analysis of biomass composition across eight developmental stages [6], carbohydrate content, starch synthesis, and metabolic enzyme activity during the vegetative growth stage [7], the distribution pattern of non-structural carbohydrates during dormancy, and alterations in the levels of soluble sugar, soluble protein, starch, glucose, fructose, and sucrose throughout the flowering phase [8] in bulbs. Furthermore, studies have examined the variation in hormone levels across different growth stages [9] and the hormone-related genes in bulbs of varying ages [10]. Physiologically, these investigations reveal that the flowering process in *L. radiata* is energy-intensive, with the bulb acting primarily as a "source" during the peak flowering period and as a "sink" throughout the extended leaf growth phase. The sink-source conversion involves a series of metabolic activities within the bulb, such as starch synthesis and breakdown, sucrose utilization, glucose accumulation, and hormone synthesis and regulation [8–10]. However, these metabolic processes have yet to be confirmed at the metabolic level, leaving the metabolic regulation mechanisms of development and flowering in *Lycoris* plants largely undefined.

Cellular signaling, energy transfer, and intercellular communication, essential to plant growth and development, are modulated by metabolites [11,12]. Metabolomics, through both qualitative and quantitative analysis of all low molecular weight metabolites, has emerged as a formidable tool for delving into the cellular biology of plants. While transcriptomic and proteomic technologies have been employed to probe the molecular underpinnings of *Lycoris* development and flowering [10,13–15], metabolomics is essential to bridge the genotype-phenotype gap. Recently, metabolomics has been applied to study the metabolic regulation mechanisms behind bulblet formation [16], petal color development [14], and galanthamine biosynthesis [17] in *Lycoris*. However, the metabolic regulation mechanisms across the different developmental stages of *Lycoris* remain unexplored. This study pioneers a comparative metabolome analysis to chart the dynamic shifts in metabolite accumulation at the bulb tips across six developmental stages of *L. radiata*, utilizing non-targeted metabolomics based on ultra-high-performance liquid chromatography quadrupole time-of-flight mass spectrometry (UHPLC-Q-TOF-MS/MS). This analysis aims to lay the groundwork for elucidating the regulatory mechanisms underpinning the development and flowering of *L. radiata*.

2. Materials and Methods

2.1. Plant Materials

From December 2018 to November 2019, a total of 48 underground bulbs of *L. radiata*, representing six distinct developmental stages, were systematically collected from Jiangxi Agricultural University Flower Gardens, Nanchang, Jiangxi Province, China. The stages included the stage of leafing out (LO, collected on 9 October), rapid leaf extension (RLE, collected on 27 October), leaf maturity (LMa, collected on 30 December), leaf withering (LWi, collected on 13 April), dormancy (Dor, collected on 28 June), and flowering (Flo, collected on 3 September). Nanchang's subtropical monsoon climate features an annual mean rainfall of 1600–1700 mm, an annual average minimum temperature of 3.9 °C, and an annual average maximum temperature of 38.7 °C. The soil is predominantly red clay with a pH value of 6.43. Bulb tips were immediately placed in liquid nitrogen upon collection and stored at −80 °C until analysis. Each developmental stage comprised eight independent biological replicates.

2.2. Sample Extraction

Freeze-dried samples were pulverized to a fine powder in liquid nitrogen. Approximately 0.06 g of the powder was dissolved in a 70% methanol aqueous solution. Subsequently, 1 mL of a pre-chilled methanol-acetonitrile solution (2:2:1, v/v) was added to 60 mg of the lyophilized powder, vortexed for 60 s, and subjected to low-temperature ultrasonication twice, each for 30 min. The sample was then centrifuged at 14,000 rcf for 20 min at 4 °C. The supernatant was collected and concentrated to dryness in preparation for UPLC-MS/MS analysis performed by Shanghai Applied Protein Technology Co., Ltd. (Shanghai, China). A quality control (QC) sample, created by combining equal volumes of extract from each developmental stage, was used to evaluate the system's stability throughout the experiment.

2.3. LC-MS Analysis

The chromatographic conditions were set with a column temperature of 25 °C and a flow rate of 0.3 mL/min. The mobile phase A consisted of water with 25 mM ammonium acetate and 25 mM ammonia, while mobile phase B was acetonitrile. A linear gradient of mobile phase B was employed as follows by volume ratio: 0–0.5 min, 95% B; 0.5–7 min, 95–65% B; 7–8 min, 65–40% B; 8–9 min, 40% B; 9–9.1 min, 40–95% B; and 9.1–12 min, 95% B. To mitigate the impact of instrument detection signal fluctuations, samples were analyzed in a randomized sequence. Mix all samples in equal volumes to form Quality Control (QC) samples. The QC samples were interspersed within the sample queue to monitor and assess the system's stability and the reliability of the experimental data.

Electrospray ionization (ESI) in both positive and negative ion modes was utilized to detect the mass spectrum signals of the samples. Following detection, metabolites were identified using an AB Triple TOF 6600 mass spectrometer, and primary and secondary mass spectra of the QC samples were collected.

2.4. Statistical Analysis

The original data were converted into mzXML format using ProteoWizard, followed by peak alignment, retention time correction, and peak area extraction via the XCMS program. Metabolite structures were identified by matching the accurate mass number (<25 ppm) and secondary spectrum. The R software package ropls (Version 1.6.2) facilitated principal component analysis (PCA) and orthogonal least squares discriminant analysis (OPLS-DA), employing seven cycles of interactive validation to assess model stability. The variable importance for projection (VIP) score derived from the OPLS-DA model identified metabolites as differential (DAMs) if they exhibited a VIP score $\geq$1 and a t-test p-value ≤ 0.05. Pathway enrichment analysis of the DAMs was performed using the Kyoto Encyclopedia of Genes and Genomes (KEGG) database, with a significance threshold of q-value ≤ 0.05 (p-value < 0.05 after false discovery rate correction).

3. Results

3.1. PCA of the Metabolites in Lycoris Bulb Samples

XCMS software (released in 2006) was used to extract the ion peaks of metabolites. The number of ion peaks in positive ion mode was 7460, and the number of ion peaks in negative ion mode was 8074. Principal component analysis (PCA) scores showed that the QC samples were closely clustered together, and the metabolomic data of the same sample, with at least five biological repetitions, were also clustered together (Figure 1). This indicates that the experiment had good repeatability. In addition, the samples of Dor, RLE, LMa, and LWi at four developmental stages were clearly separated from each other and were distinct from Flo and LO (Figure 1). However, there was no obvious separation between Flo and LO, indicating that the metabolites of Dor, RLE, LMa, and LWi at four developmental stages of *L. radiata* were different, whereas the differences between the metabolites of Flo and LO were small (Figure 1).

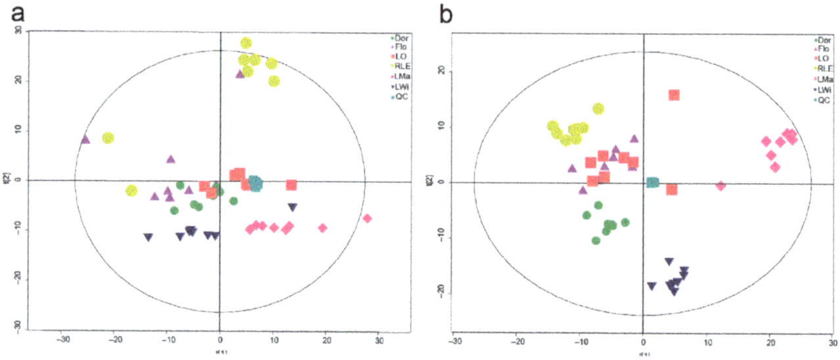

Figure 1. Principal component analysis (PCA) scores for *Lycoris* bulb samples: (**a**) PCA scores in positive ion mode (POS); (**b**) PCA scores in negative (NEG) ion mode. Dor—dormancy; Flo—lowering; LO—leafing out; RLE—rapid leaf extension; LMa—leaf maturity; LWi—leaf withering; QC—quality control.

3.2. *Orthogonal Projections to Latent Structures-Discrimination Analysis (OPLS-DA)*

The OPLS-DA model was used to analyze the data and further explore the differences between adjacent developmental stages. The samples from two adjacent developmental stages were located in different quadrants on the left and right sides of the Y-axis (Figure 2). In the positive mode, except for the RLE vs. LO group, model evaluation parameters (R2X, R2Y) and prediction ability (Q2) were all greater than 0.5 (Figure 2a1–f1). In the negative mode, R2X, R2Y, and Q2 were all greater than 0.35 (Figure 2a2–f2). This indicated that the OPLS-DA model was stable and highly reliable for the dataset and could be used to describe the differences between different developmental stages of *L. radiata*. The permutation test was used to further verify the reliability of the OPLS-DA model. All Q2 points from left to right are lower than the original Q2 points on the rightmost side, indicating that the model is robust and reliable without overfitting (Figure S1).

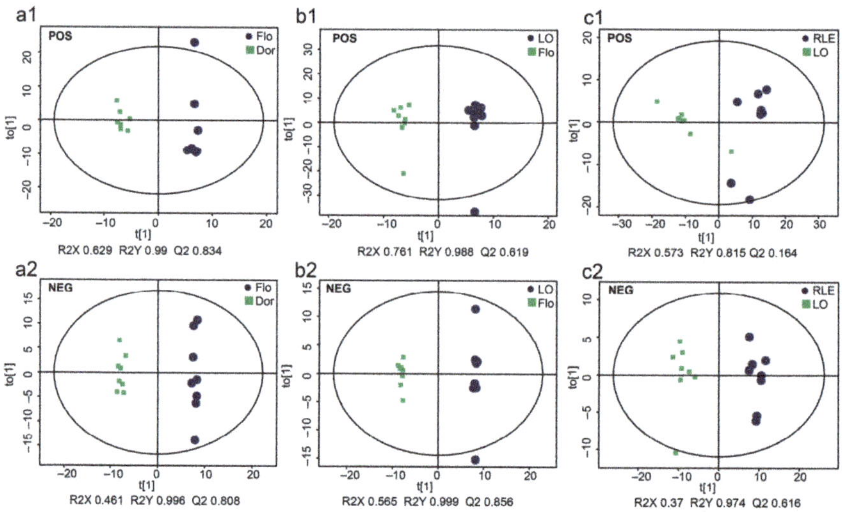

Figure 2. *Cont.*

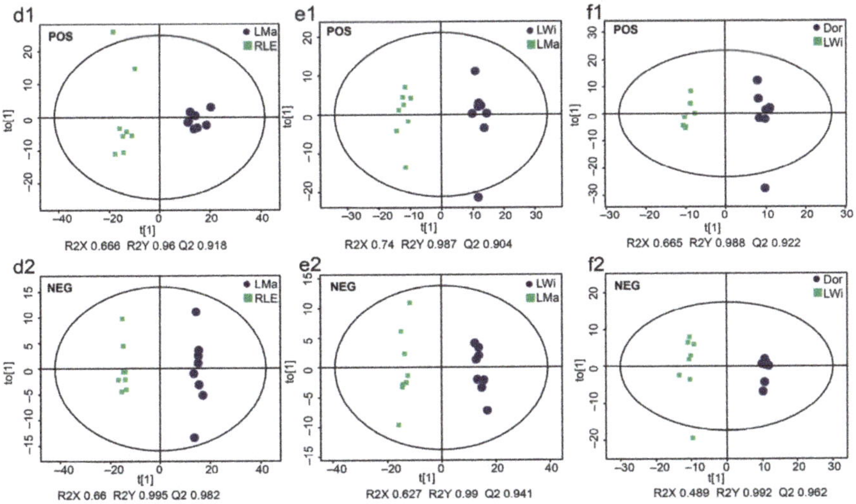

Figure 2. OPLS-DA score plots for six comparison groups in positive (**a1–f1**) and negative (**a2–f2**) ion modes. (**a1,a2**) Flo vs. Dor; (**b1,b2**) LO vs. Flo; (**c1,c2**) RLE vs. LO; (**d1,d2**) LMa vs. RLE; (**e1,e2**) LWi vs. LMa; (**f1,f2**) Dor vs. LWi. Dor—dormancy; Flo—flowering; LO—leafing out; RLE—rapid leaf extension; LMa—leaf maturity; LWi—leaf withering; POS—positive; NEG—negative.

3.3. Differentially Accumulated Metabolites (DAMs) Analysis

Based on the OPLS-DA model, DAMs between two adjacent developmental stages of *L. radiata* were identified with a VIP score ≥ 1 and a *t*-test *p*-value < 0.05. In the positive ion mode, 195 DAMs were identified, comprising 101 over-accumulated and 94 down-accumulated metabolites. Conversely, in the negative ion mode, 315 DAMs were identified, including 146 over-accumulated and 169 down-accumulated metabolites (Figure 3, Table S1). In the positive ion mode, the comparison group with the highest number of DAMs was LMa vs. RLE (41), followed by LWi vs. LMa (39), with the lowest being Flo vs. Dor (24) (Figure 3a). The most over-accumulated DAMs were observed in the LO vs. Flo comparison group (24), with LWi vs. LMa (21) following. The fewest over-accumulated DAMs occurred in the RLE vs. LO comparison (only 7). For down-accumulated DAMs, LMa vs. RLE had the most (23), followed by RLE vs. LO (21), with the fewest in LO vs. Flo (7). In the negative ion mode, the highest number of DAMs was identified in the RLE vs. LO (68) and LMa vs. RLE (68) comparison groups, with the fewest in the Flo vs. Dor group (36) (Figure 3b). The most over-accumulated metabolites were found in LO vs. Flo (36), followed by LMa vs. RLE (34), and the fewest over-accumulated metabolites were in the Dor vs. LWi comparison (10). Regarding down-accumulated DAMs, RLE vs. LO showed the most (47), followed by LMa vs. RLE (34), and the fewest were in LO vs. Flo (10).

We further analyzed the proportion of DAMs within each subclass relative to the total DAMs in each comparison group. Figure 4a reveals that the majority of DAMs across all six comparison groups belonged to the subclasses "carbohydrates and carbohydrate conjugates" and "amino acids, peptides, and analogs". Additionally, the proportion of DAMs categorized as "fatty acids and conjugates", "pyridine carboxaldehydes", "pyrimidine nucleotide sugars", "quaternary ammonium salts", and "tricarboxylic acids and derivatives" in the Flo vs. Dor comparison was higher than in other groups. The DAMs in the subclasses "beta hydroxy acids and derivatives", "linoleic acids and derivatives", and "pyrimidines and pyrimidine derivatives" were more prevalent in the LO vs. Flo comparison compared to others (Figure 4a). A radar chart analysis (Figure 4b) showed that the number of over-accumulated DAMs in the "carbohydrates and carbohydrate conjugates" subclass was notably higher in the LMa vs. RLE and RLE vs. LO comparisons than in the other four groups.

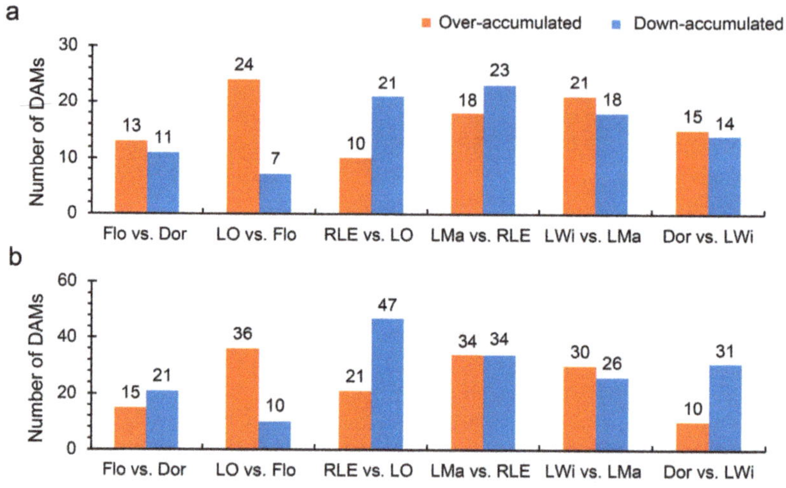

Figure 3. Number of differentially accumulated metabolites (DAMs) identified in *Lycoris* bulb Samples. (**a**) in positive (POS) ion mode and (**b**) in negative (NEG) ion mode. Dor—dormancy; Flo—flowering; LO—leafing out; RLE—rapid leaf extension; LMa—leaf maturity; LWi—leaf withering.

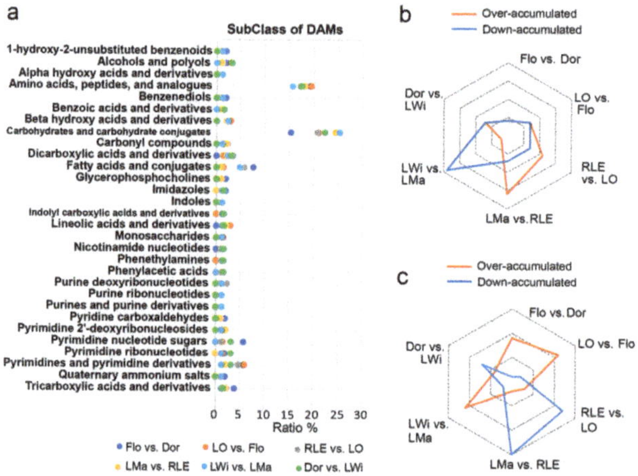

Figure 4. Subclass distribution of DAMs identified under positive and negative ion modes. (**a**) The ratio of the number of DAMs per subclass to the total DAMs within each comparison group. (**b**) Number of DAMs in the carbohydrates and carbohydrate conjugates subclasses across comparison groups. (**c**) Number of DAMs in the amino acids, peptides, and analogs subclasses across comparison groups. Dor—dormancy; Flo—flowering; LO—leafing out; RLE—rapid leaf extension; LMa—leaf maturity; LWi—leaf withering.

Additionally, the number of down-accumulated DAMs was highest in the LWi vs. LMa comparison. Meanwhile, the "amino acids, peptides, and analogs" subclass exhibited the most over-accumulated DAMs in LWi vs. LMa, followed by the LO vs. Flo and Flo vs. Dor comparisons. This subclass also exhibited the most down-accumulated DAMs in LMa vs. RLE and RLE vs. LO comparisons (Figure 4c).

3.4. The KEGG Annotation Analysis of DAMs

KEGG pathway enrichment analysis was performed based on the DAMs identified under positive and negative ion modes. The DAMs of the six comparison groups were significantly enriched in 59 (Flo vs. Dor), 45 (LO vs. Flo), 63 (RLE vs. LO), 56 (LMa vs. RLE), 55 (LWi vs. LMa), and 47 (Dor vs. LWi) metabolic pathways (Table S2) (p-value < 0.05). The top 20 KEGG pathways for each of the six comparison groups are illustrated in Figure 5, with the significance of key pathways determined by a q-value. For all comparison groups, DAMs were primarily associated with pathways such as "ABC transporters", "protein digestion and absorption", "aminoacyl-tRNA biosynthesis", "alanine, aspartate and glutamate metabolism", and "galactose metabolism", among others.

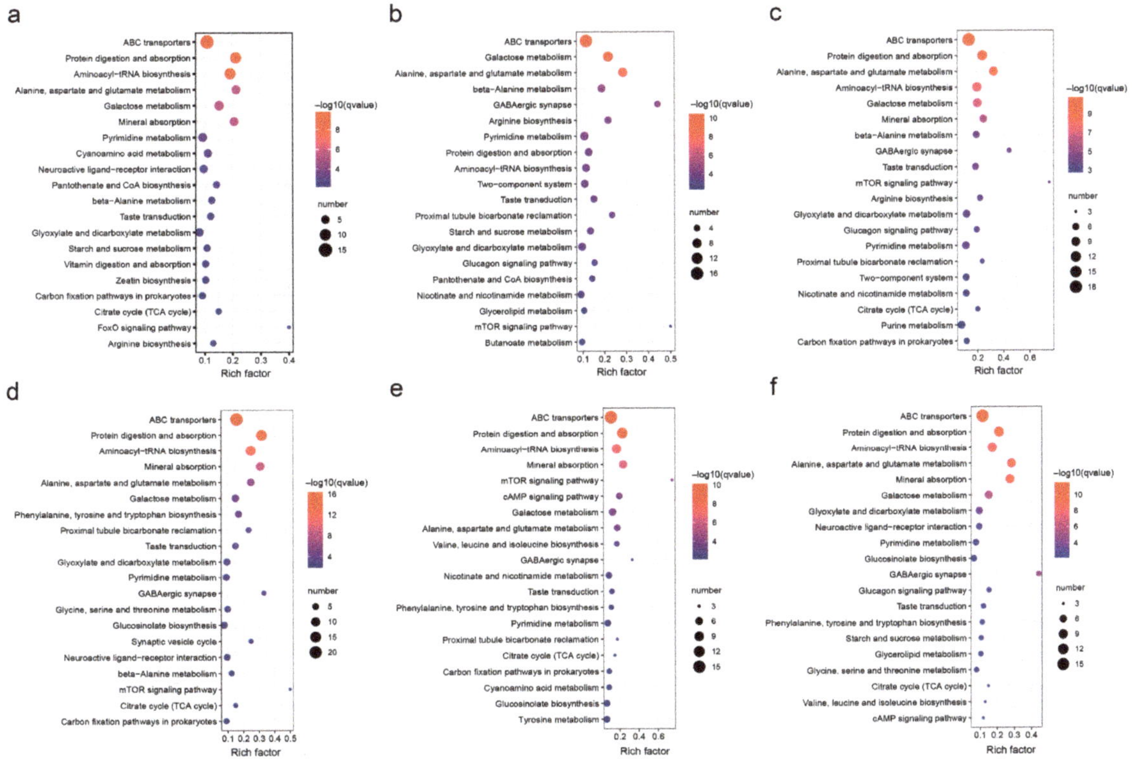

Figure 5. Top 20 KEGG enrichment of DAMs for six comparison groups. (**a**) Flo vs. Dor, (**b**) LO vs. Flo, (**c**) RLE vs. LO, (**d**) LMa vs. RLE, (**e**) LWi vs. LMa, (**f**) Dor vs. LWi.

The enriched pathways of DAMs across the six comparison groups were mainly categorized into five types: amino acid metabolism, carbohydrate metabolism, signal transduction, energy metabolism, and biosynthesis of other secondary metabolites (Figure S2). A significant proportion of the metabolic pathways were related to amino acid and carbohydrate metabolism. Specifically, the percentage of pathways enriched by DAMs related to carbohydrate metabolism in the LO vs. Flo comparison was the highest among all groups, followed by RLE vs. LO. The proportion of pathways related to amino acid metabolism enriched by DAMs in LMa vs. RLE was highest, followed by Dor vs. LWi. Additionally, the proportion of pathways associated with signal transduction and the biosynthesis of other secondary metabolites enriched by DAMs in the Flo vs. Dor comparison exceeded those in other groups, followed by LWi vs. LMa.

To elucidate developmental stage differences in *L. radiata*, we conducted a statistical analysis of KEGG pathways involved in amino acid and carbohydrate metabolism across each comparison group. The pathways associated with carbohydrate metabolism enriched by DAMs across the groups included "galactose metabolism", "glyoxylate and dicarboxylate metabolism", "the citrate cycle", and "starch and sucrose metabolism". Notably, "C5-branched dibasic acid metabolism" was uniquely enriched by DAMs in the Flo vs. Dor comparison, "glycolysis/gluconeogenesis" was specific to the LO vs. Flo comparison, and the "pentose phosphate pathway" was distinctive to the LMa vs. RLE comparison (Figure 6a). In terms of amino acid metabolism, the pathways enriched by DAMs across the groups included "alanine, aspartate, and glutamate metabolism", "arginine biosynthesis", and "glycine, serine, and threonine metabolism". "Lysine biosynthesis" was specifically enriched by DAMs in the Flo vs. Dor comparison (Figure 6b).

a

Carbohydrate metabolism	Flo vs. Dor	LO vs. Flo	RLE vs. LO	LMa vs. RLE	LWi vs. LMa	Dor vs. LWi
Amino sugar and nucleotide sugar metabolism	5	5	5	0	0	5
Ascorbate and aldarate metabolism	0	0	3	3	0	0
Butanoate metabolism	0	4	4	3	3	3
C5-Branched dibasic acid metabolism	3	0	0	0	0	0
Citrate cycle (TCA cycle)	3	3	4	3	3	3
Fructose and mannose metabolism	0	3	3	4	4	0
Galactose metabolism	7	10	9	7	6	7
Glycolysis / Gluconeogenesis	0	2	0	0	0	0
Glyoxylate and dicarboxylate metabolism	5	6	7	6	4	6
Pentose phosphate pathway	0	0	0	3	0	0
Propanoate metabolism	0	3	4	0	3	0
Pyruvate metabolism	0	3	3	0	3	0
Starch and sucrose metabolism	4	5	4	3	3	4

b

Amino acid metabolism	Flo vs. Dor	LO vs. Flo	RLE vs. LO	LMa vs. RLE	LWi vs. LMa	Dor vs. LWi
Alanine, aspartate and glutamate metabolism	6	8	9	7	5	8
Arginine and proline metabolism	0	4	4	4	0	3
Arginine biosynthesis	3	5	5	3	2	2
Cysteine and methionine metabolism	4	0	4	3	0	0
Glycine, serine and threonine metabolism	3	4	5	5	4	4
Lysine biosynthesis	2	0	0	0	0	0
Lysine degradation	0	0	0	3	0	3
Phenylalanine metabolism	0	0	4	3	4	4
Phenylalanine, tyrosine and tryptophan biosynthesis	3	0	3	6	4	4
Tyrosine metabolism	4	0	0	5	5	0
Valine, leucine and isoleucine biosynthesis	3	0	2	3	4	3

Figure 6. KEGG pathways associated with carbohydrate (**a**) and amino acid (**b**) metabolism across comparison groups, displaying DEM counts in each box. The color changes from yellow to red, with a closer proximity to red indicating a higher number of DAMs, a red box signifies a higher number of DAMs, while a blue box indicates the absence of DAMs. Dor—dormancy; Flo—flowering; LO—leafing out; RLE—rapid leaf extension; LMa—leaf maturity; LWi—leaf withering.

4. Discussion

4.1. Stress Responses during Dormancy and Flowering Stages

The dormancy stage (Dor) of *L. radiata*, spanning from May to July, is characterized by the absence of aboveground leaves while the underground bulbs undergo complex flower bud differentiation, involving the development of stages, ovaries, pistils, anthers, and pollen [6]. Following this period, the plant transitions into the late flower development stage from mid to late July. During these critical phases, our study has identified significant metabolic changes that underscore the plant's adaptation to its growth and environmental conditions.

Our results indicate a notable increase in the proportion of DAMs associated with "tricarboxylic acids and derivatives" and "quaternary ammonium salt" during the Flo vs. Dor comparison. The tricarboxylic acid (TCA) cycle is essential not only for energy production but also for integrating various biochemical processes [18]. It plays a significant role in the oxidative defense mechanism, maintaining a balance with reactive oxygen species (ROS) and contributing to cellular immunity under stress [19]. Tricarboxylic acids and their derivatives play a core role in this process [20]. Quaternary ammonium compounds

(QACs) are vital for osmotic adjustment and response to stress in plants [21], with studies demonstrating their effectiveness in alleviating drought stress in maize, which results in higher grain yields [22]. Moreover, this study highlights the unique enrichment of "C5 branched chain dicarboxylic acid metabolism" in the Flo vs. Dor comparison. This pathway is implicated in various plant stress responses, including drought stress in *Phaseolus vulgaris* and maize and cadmium stress tolerance in *Brassica juncea* [23–25]. It also plays a critical role in reprogramming protoplasts to stem cells during cell division in *Physcomitrella patens* [26]. *Lycoris* species are sensitive to environmental changes, particularly temperature variations during dormancy, which significantly affect flower bud differentiation, flowering period, and flowering rate [5]. *L. radiata* experienced temperature stimuli, stress responses, and immune reactions during its dormancy and flowering stages [13]. The metabolomic data from this study indicate specialized roles for tricarboxylic acids and derivatives, QACs and C5 branched chain dicarboxylic acid metabolism in managing stress responses during crucial developmental transitions such as flower bud differentiation and flowering of *L. radiata*.

4.2. Active Carbohydrate Metabolism during the Vegetative Growth Stages

Carbohydrate metabolism is crucial for the energy supply and biosynthetic processes in plants. In *L. radiata*, non-structural carbohydrates (NSCs), including starch and soluble sugars, play pivotal roles in bulb enlargement, leaf growth, and flower bud differentiation [27]. These carbohydrates are the bulbs' key storage, metabolic, and regulatory agents [28,29]. During the nutritional growth stages—leafing out (LO), rapid leaf extension (RLE), and leaf maturity (LMa)—the bulb functions as a repository where carbohydrates synthesized by the leaves are transported and stored [5,6]. Investigations into NSC fluctuations throughout *L. radiata*'s annual growth cycle revealed that the aggregate content of NSCs, including soluble sugars, fructose, and reducing sugars, peaked during the LMa stage.

In contrast, the accumulation of starch and sucrose was delayed, reaching peak levels during the leaf withering (LWi) stage [27]. Previous transcriptomic analyses corroborate these findings, showing significant upregulation of genes associated with starch and sucrose metabolism, and overall carbohydrate metabolism during leaf development in *L. radiata* [15]. During the RLE and LMa stage of *L. radiata*, differentially expressed genes (DEGs) and DAMs were both enriched in several carbohydrate metabolism pathways, including glyoxylate and dicarboxylate metabolism, citrate cycle (TCA cycle), starch and sucrose metabolism, and butanoate metabolism (Figure S3). Consistently, the present study observed an elevated proportion of over-accumulated metabolites in the carbohydrate and carbohydrate conjugate subclasses during the RLE vs. LO and LMa vs. RLE comparisons, indicating a more active carbohydrate metabolism at the bulb tips during the vegetative growth stages compared to the reproductive stage.

Moreover, this study identified specific enhancements in pathways associated with carbohydrate degradation: glycolysis/gluconeogenesis and the pentose phosphate pathway were notably enriched during the LO vs. Flo and LMa vs. RLE stages, respectively. Glycolysis is a central metabolic pathway converting glucose into pyruvate, which generates energy in the form of ATP and NADH, whereas gluconeogenesis involves the formation of glucose from non-carbohydrate sources [30]. The pronounced activity of glycolysis/gluconeogenesis during the LO vs. Flo comparison may reflect a shift in energy requirements as *L. radiata* transitions from reproductive to nutritional growth. Conversely, the pentose phosphate pathway, crucial for producing NADPH and ribose 5-phosphate—essential for biosynthetic reactions and maintaining cellular redox balance—was specifically enriched during the LMa vs. RLE stage [31]. Ribose 5-phosphate also serves as a precursor for synthesizing nucleotides and nucleic acids [32]. It suggests the importance of the glycolysis/gluconeogenesis pathway in supporting rapid cell proliferation and tissue expansion during the leaf growth period of *L. radiata*.

4.3. Robust Amino Acid Metabolism during the Reproductive Stages

Amino acid metabolism is crucial in plants, supporting many biological processes such as protein biosynthesis, nitrogen assimilation and transportation, synthesis of secondary metabolites, stress responses, growth regulation, and signal transduction [33]. This study observed that the proportion of upregulated metabolites identified as amino acids, peptides, and analogs was highest in the LWi vs. LMa stage, followed by the LO vs. Flo and Flo vs. Dor stage. The LWi stage marks the end of vegetative growth for *L. radiata*, transitioning into the developmental phases of inflorescence and floret primordial formation, including the development of both inner and outer perianths [6]. Conversely, the Dor stage is crucial for the development of stamens, ovaries, pistils, and the formation of anthers and pollen [6]. Plants typically accumulate soluble proteins prior to the morphological differentiation of flower buds [34]. Upon commencement of differentiation, these proteins are degraded and converted into various amino acids [35]. These amino acids are then utilized to construct structural proteins for new cells or synthesize diverse enzymes that aid cellular metabolism [35]. The findings of this study suggest robust amino acid metabolism during the reproductive stages of *L. radiata*, which is intricately linked to flower bud differentiation.

Lycoriaceae alkaloids have high medicinal value, primarily including lycorine, galanthamine, and lycoamine [1]. Among them, lycorine is the most representative compound, confirmed to have significant pharmacological activity and exhibiting notable anti-tumor activity in cancer cells [36,37]. In this non-targeted metabolomic study, two stereoisomers of lycorine were identified: one with higher accumulation levels in the Dor, Flo, and LWi stages, and the other with higher accumulation levels in the LWa and LWi stages (Table S3). This indicated that the synthesis of lycorine in the bulb tips of *L. radiata* may mainly occur during reproductive stages. The enrichment of amino acid metabolism pathways at the bulb tips of *L. radiata* may also be linked to the biosynthesis of lycorine [38].

Additionally, our analysis highlighted the unique enrichment of lysine biosynthesis in the Flo vs. Dor comparison group, underscoring its importance during the flowering stage of *L. radiata*. Notably, two differential metabolites (DAMs) in this pathway, aspartate and lysine, were upregulated. This aligns with findings from Park et al. [39], who observed an increase in most amino acids, including aspartate, during flower development in *L. radiata*. The activation of lysine biosynthesis by aspartate and lysine represents a critical biochemical pathway essential for the development of *L. radiata* flowers, indicating a specific metabolic focus during pivotal reproductive phases.

5. Conclusions

This study comprehensively analyzed the metabolic adaptations of *L. radiata* across various developmental stages, elucidating significant shifts in stress response mechanisms carbohydrate, and amino acid metabolism that are crucial for its growth and flowering. During the dormancy and flowering stages, *L. radiata* demonstrates significant metabolic responses to environmental stress, notably through the activation of the metabolites involved in tricarboxylic acids and derivatives, QACs and the pathway of C5 branched chain dicarboxylic acids metabolism. In the vegetative stages, there is a pronounced emphasis on carbohydrate metabolism. The increased activity in carbohydrate-related pathways, particularly glycolysis and the pentose phosphate pathway, underscores the metabolic adjustments needed during periods of rapid growth and development. Furthermore, the study highlighted a robust amino acid metabolism during reproductive phases. The specific enrichment of lysine biosynthesis during the flowering stage accentuates its importance in reproductive success. Investigating the genetic regulation underlying these metabolic pathways may offer deeper insights into the physiological and molecular mechanisms driving the growth cycles of *L. radiata*. Our comprehensive metabolomic analysis of *L. radiata* across various developmental stages provides significant insights that can be leveraged to optimize cultivation practices. The metabolic shifts observed throughout the plant's growth and dormancy stages reveal critical periods where targeted interventions can enhance growth, stress resilience, and flowering success.

Supplementary Materials: The following supporting information can be downloaded at: https://www.mdpi.com/article/10.3390/horticulturae10060636/s1, Figure S1: Permutation tests for the OPLS-DA model across six comparison groups in positive (a1–f1) and negative (a2–f2) ion modes; Figure S2: Classification of Key KEGG pathways enriched by DAMs and the proportion of pathways in each class relative to the total identified pathways; Figure S3: KEGG pathway analysis revealed the co-enrichment of differentially expressed genes (DEGs) and DAMs; Table S1: Full list of differentially accumulated metabolites; Table S2: KEGG enrichment of DAMs for six comparison groups ($p < 0.05$); Table S3: Accumulation status of lycorine in different developmental stages.

Author Contributions: Conceptualization, J.C. and X.J.; data curation, X.Y.; validation, X.W. and H.C.; writing—original draft preparation, X.J.; writing—review and editing, X.W., H.C. and J.C.; funding acquisition, J.C. and X.J. All authors have read and agreed to the published version of the manuscript.

Funding: This work was supported by the National Natural Science Foundation of China (32360420) and the Jiangxi Provincial Natural Science Foundation (20212BAB215014).

Data Availability Statement: Data are contained within the article and Supplementary Materials.

Conflicts of Interest: The authors declare no competing interests.

References

1. Cahlikova, L.; Breiterova, K.; Opletal, L. Chemistry and biological activity of alkaloids from the genus *Lycoris* (Amaryllidaceae). *Molecules* **2020**, *25*, 4797. [CrossRef] [PubMed]
2. Zhang, W.; Yang, J.; Chen, Y.; Xue, R.; Mao, Z.; Lu, W.; Jiang, Y. Lycorine hydrochloride suppresses stress-induced premature cellular senescence by stabilizing the genome of human cells. *Aging Cell* **2021**, *20*, e13307. [CrossRef] [PubMed]
3. Hong, L.; Ji, Y.; Kennelly, E.J.; Long, C.; Di, R. Amaryllidaceae alkaloids from *Lycoris* suppress amyloid β-induced neurodegeneration in transgenic Caenorhabditis elegans CL2355. *Ind. Crop. Prod.* **2022**, *189*, 115798. [CrossRef]
4. Kawano, S. *Lycoris sanguinea* Maxim. (Amaryllidaceae). *Plant Spec. Biol.* **2009**, *24*, 139–144. [CrossRef]
5. Cai, J.H.; Wei, X.Y.; Li, J.F.; Zhang, L. The regulation and control of ambient temperature on growth rhythm of *Lycoris radiata*. *Acta Agric. Univ. Jiangxiensis* **2018**, *40*, 24–31. (In Chinese) [CrossRef]
6. Cai, J.; Fan, J.; Wei, X.; Zhang, L. A three-dimensional analysis of summer dormancy in the red spider lily (*Lycoris radiata*). *Hortscience* **2019**, *54*, 1459–1464. [CrossRef]
7. Xu, J.; Li, Q.; Yang, L.; Li, X.; Wang, Z.; Yang, Z.; Zhang, Y. Changes in starch synthesis and metabolism within developing bulbs of *Lycoris radiata* during the vegetative growth stage. *J. Plant Growth Regul.* **2019**, *39*, 785–794. [CrossRef]
8. Zhang, Y.; Yang, X.; He, L.; Wang, L. Relationships between flower bud differentiation in two kinds of *Lycoris* and the changes of carbohydrate, antioxidant enzymes and endogenous hormones. *J. Anhui Agric. Univ.* **2019**, *46*, 342–349. (In Chinese) [CrossRef]
9. Fan, J.; Zhang, Y.; Wu, W.; Yan, X.; Yang, X.; Wang, L. Changes of endogenous hormones in 2 kinds of *Lyroris* plants at different growth stages. *Mol. Plant Breed.* **2022**, *20*, 6550–6558. (In Chinese) [CrossRef]
10. Zhang, F.; Cheng, G.; Shu, X.; Wang, N.; Wang, Z. Transcriptome analysis of *Lycoris chinensis* bulbs reveals flowering in the age-mediated pathway. *Biomolecules* **2022**, *12*, 899. [CrossRef]
11. Judge, A.; Dodd, M.S. Metabolism. *Essays Biochem.* **2020**, *64*, 607–647. [CrossRef] [PubMed]
12. Baker, S.A.; Rutter, J. Metabolites as signalling molecules. *Nat. Rev. Mol. Cell Biol.* **2023**, *24*, 355–374. [CrossRef] [PubMed]
13. Jiang, X.; Chen, H.; Wei, X.; Cai, J. Proteomic analysis provides an insight into the molecular mechanism of development and flowering in *Lycoris radiata*. *Acta Physiol. Plant.* **2021**, *43*, 4. [CrossRef]
14. Wang, N.; Shu, X.; Zhang, F.; Zhuang, W.; Wang, T.; Wang, Z. Comparative transcriptome analysis identifies key regulatory genes involved in anthocyanin metabolism during flower development in *Lycoris radiata*. *Front. Plant Sci.* **2021**, *12*, 761862. [CrossRef] [PubMed]
15. Jiang, X.; Wei, X.; Chen, H.; Zhang, Y.; Ma, M.; Cai, J. Transcriptome analysis of *Lycoris radiata* bulb tips at different developmental stages. *Pak. J. Bot.* **2023**, *55*, 1023–1034. [CrossRef] [PubMed]
16. Ren, Z.M.; Zhang, D.; Jiao, C.; Li, D.Q.; Wu, Y.; Wang, X.Y.; Gao, C.; Lin, Y.F.; Ruan, Y.L.; Xia, Y.P. Comparative transcriptome and metabolome analyses identified the mode of sucrose degradation as a metabolic marker for early vegetative propagation in bulbs of *Lycoris*. *Plant J.* **2022**, *112*, 115–134. [CrossRef]
17. Li, Q.; Xu, J.; Zheng, Y.; Zhang, Y.; Cai, Y. Transcriptomic and metabolomic analyses reveals that exogenous methyl jasmonate regulates galanthamine biosynthesis in *Lycoris longituba* seedlings. *Front. Plant Sci.* **2021**, *12*, 713795. [CrossRef]
18. Zhang, T.; Peng, J.T.; Klair, A.; Dickinson, A.J. Non-canonical and developmental roles of the TCA cycle in plants. *Curr. Opin. Plant Biol.* **2023**, *74*, 102382. [CrossRef] [PubMed]
19. Mailloux, R.J.; Beriault, R.; Lemire, J.; Singh, R.; Chenier, D.R.; Hamel, R.D.; Appanna, V.D. The tricarboxylic acid cycle, an ancient metabolic network with a novel twist. *PLoS ONE* **2007**, *2*, e690. [CrossRef]
20. Choi, I.; Son, H.; Baek, J.H. Tricarboxylic acid (TCA) cycle intermediates: Regulators of immune responses. *Life* **2021**, *11*, 69. [CrossRef]
21. Ashraf, M.; Harris, P.J.C. Potential biochemical indicators of salinity tolerance in plants. *Plant Sci.* **2004**, *166*, 3–16. [CrossRef]

22. Trivedi, K.; Vijay Anand, K.G.; Kubavat, D.; Patidar, R.; Ghosh, A. Drought alleviatory potential of *Kappaphycus* seaweed extract and the role of the quaternary ammonium compounds as its constituents towards imparting drought tolerance in *Zea mays* L. *J. Appl. Phycol.* **2018**, *30*, 2001–2015. [CrossRef]
23. Zenda, T.; Liu, S.; Wang, X.; Jin, H.; Liu, G.; Duan, H. Comparative proteomic and physiological analyses of two divergent maize inbred lines provide more insights into drought-stress tolerance mechanisms. *Int. J. Mol. Sci.* **2018**, *19*, 3225. [CrossRef] [PubMed]
24. Subramani, M.; Urrea, C.A.; Kalavacharla, V. Comparative analysis of untargeted metabolomics in tolerant and sensitive genotypes of common bean (*Phaseolus vulgaris* L.) seeds exposed to terminal drought stress. *Metabolites* **2022**, *12*, 944. [CrossRef] [PubMed]
25. Yang, Z.; Tan, P.; Huang, Z.; Sun, Z.; Liu, Z.; Liu, L.; Zeng, C.; Tong, J.; Yan, M. Metabolic profiles in the xylem sap of *Brassica juncea* exposed to cadmium. *Physiol. Plant* **2023**, *175*, e13886. [CrossRef] [PubMed]
26. Xiao, L.; Zhang, L.; Yang, G.; Zhu, H.; He, Y. Transcriptome of protoplasts reprogrammed into stem cells in *Physcomitrella patens*. *PLoS ONE* **2012**, *7*, e35961. [CrossRef] [PubMed]
27. Wei, X.; Zhang, Y.; Ma, M.; Jiang, X.; Chen, H.; Wu, J.; Yang, Y.; Cai, J. Changes of non-structured carbohydrate and starch metabolizing enzyme in bulbs of *Lycoris radiata* within the annual growth cycle. *J. Nanjing For. Univ. (Nat. Sci. Ed.)* **2024**, *48*, 106–114. (In Chinese) [CrossRef]
28. Kiani, Z.; Mashayekhi, K.; Mousavizadeh, S.J.; Nezhad, K.Z. Changes in carbohydrate and phytochemical content in the bulbs and leaves of ten onion accessions between two developmental stages. *Italus Hortus* **2022**, *29*, 47–55. [CrossRef]
29. Ohyama, T.N.U.J.; Ikarashi, T.; Matsubara, T.; Baba, A. Behavior of carbohydrates in mother and daughter bulbs of tulips (*Tulipa gesneriana*). *Soil Sci. Plant Nutr.* **1988**, *34*, 405–415. [CrossRef]
30. Naifeh, J.; Dimri, M.; Varacallo, M. Biochemistry, aerobic glycolysis. In *StatPearls [Internet]*; StatPearls Publishing: Treasure Island, FL, USA, 2024. Available online: https://www.ncbi.nlm.nih.gov/books/NBK470170/ (accessed on 9 April 2023).
31. Copeland, L.; Turner, J.F. The regulation of glycolysis and the pentose phosphate pathway. In *Biochemistry of Metabolism*; Davies, D.D., Ed.; Academic Press: Cambridge, MA, USA, 1987; pp. 107–128. [CrossRef]
32. Bernhardt, H.S.; Sandwick, R.K. Purine biosynthetic intermediate-containing ribose-phosphate polymers as evolutionary precursors to RNA. *J. Mol. Evol.* **2014**, *79*, 91–104. [CrossRef]
33. Kawade, K.; Tabeta, H.; Ferjani, A.; Hirai, M.Y. The roles of functional amino acids in plant growth and development. *Plant Cell Physiol.* **2023**, *64*, 1482–1493. [CrossRef] [PubMed]
34. Lin, B.; Wang, D.; Li, Z.; Liu, J.; Xu, W.; Li, X.; Huang, J. Morphological development and nutritional metabolism during floral bud differentiation in *Acca Sellowiana* (Feijoa). *Int. J. Fruit. Sci.* **2023**, *23*, 102–115. [CrossRef]
35. Xu, J.; Li, Q.; Yang, L.; Li, X.; Wang, Z.; Zhang, Y. Changes in carbohydrate metabolism and endogenous hormone regulation during bulblet initiation and development in *Lycoris radiata*. *BMC Plant Biol.* **2020**, *20*, 117–180. [CrossRef] [PubMed]
36. Lamoral-Theys, D.; Andolfi, A.; Van Goietsenoven, G.; Cimmino, A.; Le Calve, B.; Wauthoz, N.; Megalizzi, V.; Gras, T.; Bruyere, C.; Dubois, J.; et al. Lycorine, the main phenanthridine Amaryllidaceae alkaloid, exhibits significant antitumor activity in cancer cells that display resistance to proapoptotic stimuli: An investigation of structure-activity relationship and mechanistic insight. *J. Med. Chem.* **2009**, *52*, 6244–6256. [CrossRef] [PubMed]
37. Li, A.; Du, Z.; Liao, M.; Feng, Y.; Ruan, H.; Jiang, H. Discovery and characterisation of lycorine-type alkaloids in *Lycoris* spp. (Amaryllidaceae) using UHPLC-QTOF-MS. *Phytochem. Anal.* **2019**, *30*, 268–277. [CrossRef] [PubMed]
38. Tang, M.; Li, C.; Zhang, C.; Cai, Y.; Zhang, Y.; Yang, L.; Chen, M.; Zhu, F.; Li, Q.; Li, K. SWATH-MS-Based proteomics reveals the regulatory metabolism of amaryllidaceae alkaloids in three *Lycoris* species. *Int. J. Mol. Sci.* **2023**, *24*, 4495. [CrossRef]
39. Park, C.H.; Yeo, H.J.; Kim, Y.J.; Nguyen, B.V.; Park, Y.E.; Sathasivam, R.; Kim, J.K.; Park, S.U. Profiles of secondary metabolites (phenolic acids, carotenoids, anthocyanins, and galantamine) and primary metabolites (carbohydrates, amino acids, and organic acids) during flower development in *Lycoris radiata*. *Biomolecules* **2021**, *11*, 248. [CrossRef]

horticulturae

Article

Classification of Vase Life Day Based on Petal Colorimetric Data in Cut Lisianthus Using AutoML

Hye Sook Kwon and Seong Heo *

Department of Horticulture, Kongju National University, Yesan 32439, Republic of Korea
* Correspondence: heoseong@kongju.ac.kr; Tel.: +82-41-330-1222

Abstract: This study investigated the potential of petal colorimetric data to classify vase life stages in cut lisianthus flowers (*Eustoma grandiflorum*). We analyzed the changes in the petal color space over time, focusing on the b* value as an indicator of senescence due to increasing yellowing caused by copigmentation. A comparative analysis was conducted between two cultivation methods: soil (S) and hydroponic (H) cultivation. The objective was to evaluate the performance of machine learning models trained to classify vase life stages based on petal color data. Automated machine learning models exhibited better performance in H-cultivated cut flowers, effectively distinguishing days within the vase life stages from Days 1 to 14 for H cultivation. Cut flowers cultivated under S conditions showed less variation in the color space from Days 1 to 9, maintaining a relatively uniform color range. This made it more difficult to distinguish the vase life stages compared to H cultivation. These findings demonstrate that petal color metrics can serve as reliable indicators of cut flower senescence and potentially facilitate nondestructive methods for classifying vase life stages. This technology holds promise for wider applications in the floriculture industry, improving quality control, and extending the vase life of various cut-flower crops.

Keywords: CIELAB; lisianthus; machine learning; senescence; vase life

1. Introduction

The generic name "Eustoma" combines the Greek words "Eu" (beauty) and "stoma" (mouth), meaning "flower with a large and beautiful mouth" [1–3]. The color and shape of lisianthus flowers are highly valued commercially, making them popular among growers and distributors [4]. Lisianthus is native to Texas and the Colorado plateau in the United States, where it has been cultivated since 1960 [5], and cultivation for cut flower production began in the early 1980s by Japanese seed companies [1,6,7].

In recent years, the market demand for cut flowers has continued to expand, with over 4 billion dollars globally distributed [8]. The appearance of lisianthus, with its rose-like petals, is highly appealing to consumers [9,10], leading to a sharp increase in sales over the past few years [11]. From a growers' perspective, lisianthus produces many blossoms on a single stem, and its flowers and leaves are highly durable [9], making it easy to transport worldwide [12,13]. Lisianthus can be cultivated annually, and its shipping schedule can be adjusted, making it a highly valuable crop for growers [2].

The market value of cut flowers largely depends on their vase life [8]. Cut flowers experience a decrease in moisture content owing to active transpiration and low water absorption, leading to quality degradation and shortened vase life during transportation and storage [8]. Additionally, a low moisture content inhibits water transpiration through the leaves, causing secondary water stress, which further affects the vase life of cut flowers and rapidly diminishes their decorative value [14]. Cut flowers have limited vase life because senescence accelerates after flower cutting and is typically accompanied by morphological, physiological, and biochemical deterioration [15,16]. This is induced by ethylene biosynthesis as oxidative stress associated with the production of reactive oxygen species and free radical increases [17]. Ethylene also affects the gene expression and activity of degradative enzymes such as lipoxygenase, lipase, pectinase, protease, polygalacturonase, and chlorophyllase [18–21]. The postharvest longevity of cut flowers is influenced by water and nutrient loss due to increased respiration and pathogen infection. Bacteria and fungi cause stomatal blockage, hindering the water transport [22–24]. Moreover, the vase life of cut flowers is significantly affected by the cultivation environment [25]. Lisianthus is typically produced by soil cultivation via vegetative propagation. However, vegetatively propagated lisianthus seedlings have the disadvantage of inconsistent quality in terms of flowering time, plant height, and number of flowers [26]. Furthermore, continuous cropping of lisianthus in soil cultivation often leads to severe quality degradation and yield reduction [27]. Although the reasons for the continuous cropping damage are not yet clear, they have been reported to be related to inadequate soil nutrient conditions and allelopathy [28]. Therefore, hydroponic cultivation has recently been widely applied in South Korea as an alternative to soil cultivation of lisianthus [29].

Among the several techniques prolonging the vase life and quality of cut flowers, chemical treatments are widely used in floriculture industry. Cut flowers are kept in vase solutions, containing carbohydrates, germicides, plant growth regulators, ethylene inhibitors, mineral salts, and organic acids to extend flower postharvest performance [30]. The vase life can be extended by supplying soluble sugars, such as sucrose, glucose, or fructose, in preservatives [30]. The sugars act as a source of respiratory substrate, improving the absorption capacity, helping to maintain petal turgidity and prolonging the vase life of cut flowers [24]. The germicides are antimicrobial chemicals designed to prevent the growth of bacteria, fungi, and other microorganisms [31]. Organic acids lower the pH of vase solutions, inhibiting the proliferation of bacteria that block xylem vessels in the cut area and prevent the maintenance of water balance [24]. The ethylene inhibitors, aminoethoxyvinylglycine, aminooxyacetic acid, and 1-methylcyclopropene, prevent the synthesis of ethylene, which accelerates senescence. On the contrary, several plant hormones have the effect of delaying senescence. Salicylic acid (SA) has been widely used in the floriculture industry to extend the vase life of roses, gladiolus, carnation, and other cut flowers [29]. SA is known to play an important role in systemic acquired resistance to pathogens and inhibits ethlyene production by suppressing the activities of 1-aminocyclopropane-1-carboxylate (ACC) synthase and ACC oxidase [29]. Moreover, several phenolic compounds, which are induced by SA, directly affect the photosynthetic electron transport chain or indirectly influence the photosynthetic machinery by controlling stomatal conductivity [29].

In general, destructive methods have been used to identify substances present in plants or to estimate their content, which involve grinding the plants and measuring them using a spectrophotometer or high-performance liquid chromatography (HPLC). This destructive measurement approach requires significant sample quantities, substantial time, and labor investment for the measurements. However, with the recent rise of smart farms and smart agriculture, non-destructive methods are becoming more important than destructive ones. This is because measuring crop status using real-time, non-destructive methods can be widely used throughout all stages of crop production, cultivation and distribution. These systems are implemented using hyperspectral imaging technology. Hyperspectral imaging (HSI) is a technology that combines spectroscopic and imaging techniques to accurately analyze various material properties by measuring continuous wavelengths from the visible and near-infrared spectrum range [32]. The most common wavelength range used for flavonoid detection is the visible and near-infrared spectrum, typically spanning from 400 nm to 1000 nm [33]. Our research utilized a colorimeter to detect the increase in colorless or yellowish flavonols observed as senescent changes in cut flowers, determined by b* values. The CIELAB is a three-dimensional color space with three coordinates: L* (lightness, ranging from 0 (black) to 100 (white)), a* (red–green axis, with negative values toward green and positive values toward red), and b* (blue–yellow axis, with negative values toward blue and positive values toward yellow). The a* and b* axes are unbounded, but measurement software limits them to a range such as −128 to 127 [34]. Colorimeters use three filters (red, green, blue) to take wideband readings along the visible spectrum to obtain an estimate of color and a measurement wavelength range of 380–500 nm is mentioned for detecting yellow color [34]. Although the measurement range of the colorimeter we used is limited to the visible light spectrum, our previous study has demonstrated that colorimeter is capable of detecting flavonols, which can be used to estimate the vase life of cut flowers [29].

The color of leaves and flowers in plant organs serves as a phenotypic indicator of the physiological status of plants [35]. Flower petal color is an important indicator of floral quality and can be accurately measured using a colorimeter based on the CIELAB color system [36]. The petal color of lisianthus varies significantly among cultivars, and even within the same cultivar, color can change depending on the cultivation environment [37]. Additionally, the color change in cut flowers is associated with vase life, resulting in differences in the petal color space over time, which can be used to predict or distinguish vase life [29,38,39]. The physiological and biochemical properties of pigments are closely related to the senescence process of flowers, and in our previous research, we developed a model that predicts vase life based on petal color data [29]. In this study, we have aimed to measure the CIELAB values of petals on each day of vase life, determine the color space distribution of each petal over time, and develop a machine learning model to differentiate measurement days based on these data.

2. Materials and Methods

2.1. Plant Materials

One thousand lisianthus seedlings were planted in a greenhouse at the Pocheon Agricultural Technology Center on 22 June 2022. Four cultivars were planted in equal proportions: Arena Green (AG), Blue Picote (BP), Corelli Pink (CP), and Kroma White (KW). These sixty-five-day-old seedlings were randomly divided into soil (S) or hydroponic (H) cultivation groups, and 48 seedlings were randomly selected and assigned to the two treatment groups after cultivation. This process was repeated thrice for each group. Throughout the cultivation period, the greenhouse maintained an average temperature of 28 °C and an average duration of 13.5 h, utilizing natural daylight with shading from 11:00 AM to 2:00 PM on days of high light intensity or artificial light on cloudy days. The greenhouse maintained a humidity level of 65 ± 10% and a CO_2 concentration of 393.3 ppm. Designated beds were established within the greenhouse for the cultivation of S and H.

The soil for S cultivation was prepared by mixing peat moss, horticultural substrates, perlite, and oil cake in the following proportions: 0.50:0.38:0.10:0.02. The composite soil samples were analyzed as follows: pH (1:5), 6.9; organic matter (OM), 54 g kg^{-1}; available phosphate (Av. P_2O_4), 790 mg kg^{-1}; exchangeable (Exch.) K, 14.8 cmol$_c$ kg^{-1}; Exch. Ca, 14.8 cmol$_c$ kg^{-1}; Exch. Mg, 4.8 cmol$_c$ kg^{-1}; and electrical conductivity (EC), 2.3 dS m^{-1}.

The nutrient solution for H cultivation, based on a volume of 1000 L, was prepared for tank A by combining KNO_3 (303 g), $CaNO_3$ (944 g), and Fe-EDTA (22.62 mg) and for tank B by combining KNO_3 (303 g), $NH_4H_2PO_4$ (115 g), $MgSO_4{\cdot}7H_2O$ (492 g), $CuSO_4{\cdot}5H_2O$ (78.58 mg), H_3BO_3 (2858.5 mg), $MnSO_4{\cdot}H_2O$ (1538 mg), $ZnSO_4{\cdot}7H_2O$ (219.8 mg), and $NH_4Mo_7O_{24}{\cdot}4H_2O$ (121.3 mg). The prepared nutrient solution was applied using a drip irrigation system. The nutrient solution was analyzed as follows: pH (1:5), 6.7; OM, 54 g kg^{-1}; Av. P_2O_4, 674 mg kg^{-1}; Exch. K, 13.7 cmol$_c$ kg^{-1}; Exch. Ca, 13.7 cmol$_c$ kg^{-1}; Exch. Mg, 4.4 cmol$_c$ kg^{-1}; and EC, 2.7 dS m^{-1}. The medium comprised peat moss and perlite in a 1:1 ratio.

The two treatment groups were cultivated in the same greenhouse; however, the moisture and nutrients supplied to the soil differed. In the S cultivation, lisianthus was planted three times in a container of 35 × 30 × 24 cm. For the first two weeks after planting, 2 L of water was supplied four times a week per container, and from the third week onward, the same amount of water was provided three times a week. Watering was consistently carried out between 9:00 AM and 10:00 AM.

In contrast, in the H cultivation, lisianthus was replanted in a container measuring 30 × 40 × 2500 cm filled with media. During the first two weeks after planting, only tap water was provided. In the second week, 30 mL of the nutrient solution was supplied five times a day using a drip irrigation system, each time lasting 10 s. From the fourth week onwards, 40 mL of the nutrient solution was supplied eight times a day, each time lasting 10 s. As a result, lisianthus in S cultivation occupied a smaller area than that in H cultivation, and the two treatment groups had different moisture and nutrient supplies. The time to first flowering was 56 and 58 days for S and H cultivation, respectively. In addition, the time required to harvest three flowers from each lisianthus plant was longer under H cultivation than under S cultivation.

2.2. Preparation of Cut Flowers and Collection of Colorimetric Data According to the Stages of Vase Life

The flowers were cut on 31 August 2022 for the S cultivation and on 16 September 2022 for the H cultivation. Taking into account factors such as flower size and flowering opening (whether the flower was 60–70% open), we selected and cut flowers that were at a similar stage of flowering. The four lisianthus cultivars showed differences in their flowers, stems, and the appearance and location of their leaves. The flower stems were cut to 5 cm to match the stem length of the four cultivars, and all lower leaves were removed. Cut flowers cultivated under S cultivation were measured for petal color using a colorimeter on the day of cutting (designated as Day 1) and subsequently at Days 5 and 9 after cutting. For cut flowers in H cultivation, which possess a longer vase life compared to those from S cultivation, petal color measurements were obtained at Days 1, 9, and 14. Forty-eight cut flowers were collected for each cultivar, and colorimetric data were measured in five replicates per cut flower. Thus, a total of 960 measurements were obtained for each cultivation group. Cut flowers were grown in a laboratory at a constant temperature of 20.4 °C, relative humidity of 60–76%, and illumination of 7.43–9.45 μmol m^2 s^{-1}. The water in the vase was obtained from tap water in the laboratory and replaced with fresh water every 5 days. The water levels in each vase were maintained at a constant level and no preservatives were added to ensure that the vase life of all flowers was assessed fairly. Three stems of each lisianthus species were placed in each vase. The CIELAB color space was measured using a colorimeter (CR-300; Konica Minolta, Tokyo, Japan) and consisted of L* (lightness), a*, and b* (chromaticity indices) values. Color changes were observed from the initial flowering stage of cut flowers (Day 1) until obvious wilting (Day 9 for S cultivation or Day 14 for H cultivation).

2.3. Statistical Analysis

We used a two-way interaction between cultivar and group to explain the CIELAB values (L*, a*, b*). The figures were visualized using R 4.4.1 [40] on RStudio (v. 2024.04.02) using the following packages: dplyr, emmeans, ggplot2, ggpubr, rstatix, and tidyverse. To describe senescence over time after flower cutting, we visualized the color changes of the flowers for the measurement days. For graphical demonstration, flowers grown in the S treatment group were grouped by measurement day (Day 1, 5, and 9) and cultivar (AG, BP, CP, and KW). The flowers treated in the H group were grouped by measurement day (Day 1, 9, and 14) and cultivar.

Each data point (L*, a*, b*) was plotted in 3-dimensional space where the x-, y-, and z-axes represent the values of a*, b*, and L*, respectively. The L*, a*, and b* values were converted to standardized RGB values, and each data point was colored using the associated RGB values. The 3-dimensional mean vector and covariance matrix were estimated for each cultivar and measurement day, and an ellipsoid was superimposed in the 3-dimensional space to represent a confidence region for the three expected color parameters at a confidence level of 0.95. The plot3D and rgl packages were used in R for graphical representation.

In addition, we used machine learning (ML) algorithms to classify the measurement days of vase life using petal colorimetric values. The vase life of cut flowers in soil cultivation was approximately 10 days, and measurements were obtained on Days 1, 5, and 9 after flower cutting, with five repeated measurements in 48 samples. The vase life of cut flowers in hydroponic cultivation was approximately 15 days, and measurements were obtained on Days 1, 9, and 14 in the same way as described above. The H_2O package (https://docs.h2o.ai, accessed on 1 July 2024) was used to implement the ML algorithms in R [41]. The area under the ROC curve (AUC) was calculated for classification.

ML analysis suffers from repetitive processes such as data pre-processing, variable selection, algorithm selection, training, and hyperparameter tuning. This leads to researchers from various fields experiencing considerable difficulty in applying machine learning. To facilitate this, researchers can readily develop ML models owing to the automation of processes through automated machine learning (AutoML). While numerous AutoML frameworks have been developed to date, we opted for the use of H_2O's AutoML platform [41]. We employed an ensemble method that combines various ML models to distinguish different stages of vase life. Ensemble machine learning models use multiple algorithms to obtain better classification performance than could be obtained from any of the constituent learning algorithms [41]. We used distributed random forests (DRF), gradient boosting machines (GBM), and generalized linear models (GLM), among various algorithms.

3. Results

3.1. The Changes in Petal Color Depending on Cultivation Method

In South Korea, lisianthus is primarily cultivated through soil cultivation; however, hydroponic cultivation is increasingly used to extend vase life (Figure 1). In soil cultivation, the vase life is approximately 10 days, whereas in hydroponic cultivation, it reaches approximately 15 days. After flower cutting, changes in the color and shape of the petals occur over time. The circular flower corolla becomes more distorted as the senescence progresses. Additionally, the vibrant color of the petals gradually fades as vase life increases. The main reason for this is that, as shown in Figure 1, the color of the petals turns more yellow as senescence progresses. This coloration differs from that produced by carotenoids and, when compared to other petal colors, appears as a nearly colorless or yellowish hue. This coloration arises from the expression of flavones or flavonols, which are types of flavonoids distinct from anthocyanins. These compounds act as copigments. Copigmentation is the phenomenon in which anthocyanins form complexes with copigments, such as flavonols or flavones, resulting in an enhancement of pigmentation [29]. In the case of BP, a deepening of the purple color is observed as senescence progresses, due to increased binding between anthocyanins responsible for the initial violet hue and flavones which intensify the coloration. However, in AG and KW, the expression of anthocyanins is relatively low. Consequently, as senescence progresses, the pale-yellow color of the copigments becomes more prominent. Therefore, the increased expression of these copigments is a phenotype directly linked to senescence, and an index reflecting the relative increase in their content could provide a basis for quantifying flower senescence.

The petal color of the four cultivars was measured using a colorimeter according to the day of vase life (Figure 2). In the CIELAB color space, the L* value represents the lightness: black at 0 and white at 100. The a* value represents the green–red opponent colors, and negative values toward green and positive values toward red. The b* value represents the blue–yellow opponent colors, and negative values toward blue and positive values toward yellow.

Figure 1. The morphological and color changes by cultivar, and vase life of cut lisianthus according to the cultivation methods. (**A**) Soil (S) cultivation, (**B**) hydroponic, (H) cultivation.

Across all cultivation groups, significant differences in L*, a*, and b* values were observed among the four cultivars according to their stage of senescence. In the S group, BP had an average a* value of 10.4 on Day 9, which was significantly lower than the values seen on Days 1 and 5 (13.5 and 14.2, respectively, *p*-values: 1.93×10^{-4} for Day 1 vs. Day 9, and 1.99×10^{-4} for Day 5 vs. Day 9). The average b* value on Day 9 was −7.27, which also showed a significant difference from the value on Day 5 (−10.7, *p*-value 3.63×10^{-4}) and Day 1 (−10.3, *p*-value 1.21×10^{-4}). In other words, for BP of the S group, there was a rapid change in the chromaticity indices (a* and b* values), excluding lightness (L* value), on Day 9 as the color senescence progressed. For CP in the S group, the average a* value on Day 1 was 12.3, but it significantly decreased to 8.7 on Day 9 (8.80×10^{-6}). In other words, the pink color of CP gradually fades with petal senescence. On Day 1, the average b* value for CP was 2.8; however, it increased to 4.6 on Day 9 with a statistically non-significant change. This indicates that the petal color of CP became more yellow on Day 9, which was the vase life limit for the S group. Within the S group cultivars, no significant difference in L* values was observed as senescence progressed.

Regarding the L* value, the BP cultivated in the H group showed a significant difference in L* values between Days 1 and 14 (*p*-value 7.84×10^{-13}) and between Days 9 and 14 (*p*-value 1.14×10^{-13}), unlike the other cultivars. As the vase life reached 15 days, the L* value rapidly decreased, indicating a reduction in lightness, as shown in Figure 1. In the H group, there was a trend of decreasing average L* values with increasing vase life, although the difference was not statistically significant except for BP. However, the BP in the S group showed no difference in the L* value between Days 1, 5, and 9. On Day 14, the average a* value for BP increased, making the red color more intense. The b* value decreased to −3.58 on Day 14 (*p*-value 8.22×10^{-3} for Day 9 vs. Day 14) from −0.64 on Day 9.

For the CP and KW in the H group, only the b* value changed significantly as senescence progressed. The average b* value was 4.4 on Day 1, but it increased significantly to 8.5 on Day 9 and 10.9 on Day 14 (*p*-value 2.71×10^{-4} and 1.70×10^{-12}, respectively). KW also showed a significant increase only in the b* value, similar to CP. The average b* value, which showed no difference between Day 1 and Day 9 (4.4 and 3.9, respectively), suddenly increased to 9.1 on Day 14 (*p*-value 2.01×10^{-7} for Day 1 vs. Day 14, and 2.27×10^{-6} for Day 9 vs. Day 14). Unlike BP, CP and KW showed an increase in the b* value by Day 14, indicating an increase in copigment expression. As petal senescence progressed significantly, the expression of the yellow color increased. Although senescence-induced changes in b* values varied across cultivars and cultivation method, there was a general tendency for b* values to increase as senescence progressed.

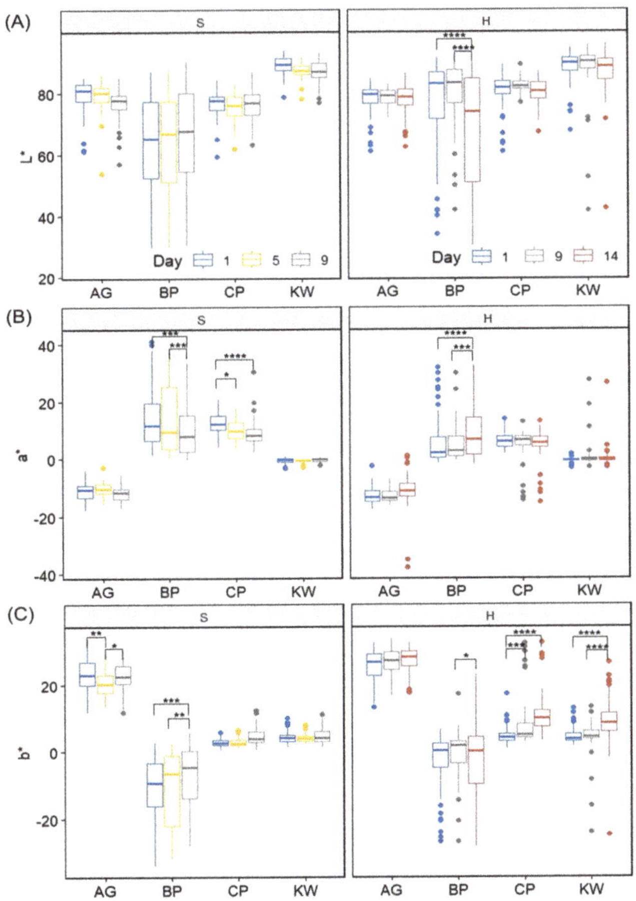

Figure 2. Two-way ANOVA results examining the influence of cultivation method and measurement day for vase life on the CIELAB color values in four lisianthus cultivars: (**A**) L* value, (**B**) a* value, and (**C**) b* value. Multiple pairwise comparisons were applied to perform post hoc analysis using the 'emmeans' package in R. The statistical significance is indicated by * ($p < 0.05$), ** ($p < 0.01$), *** ($p < 0.001$), and **** ($p < 0.0001$). Not significant, ns has not been indicated.

3.2. Three-Dimensional Plot Based on Petal Colorimetric Data

The three-dimensional color space was arranged according to the cultivar and measurement day (Figure 3 for the S group and Figure 4 for the H group). The points in the figure represent the petal color obtained by converting the CIELAB values into the RGB color. AG and KW formed an elongated ellipse parallel to the b* axis. This means that the variation in the b* values was large when compared to the other values. In contrast, CP formed an ellipse along the a* axis, indicating pink petals with a high a* value. BP had a purple petal color and exhibited the characteristics of a large a* value and a small b* value.

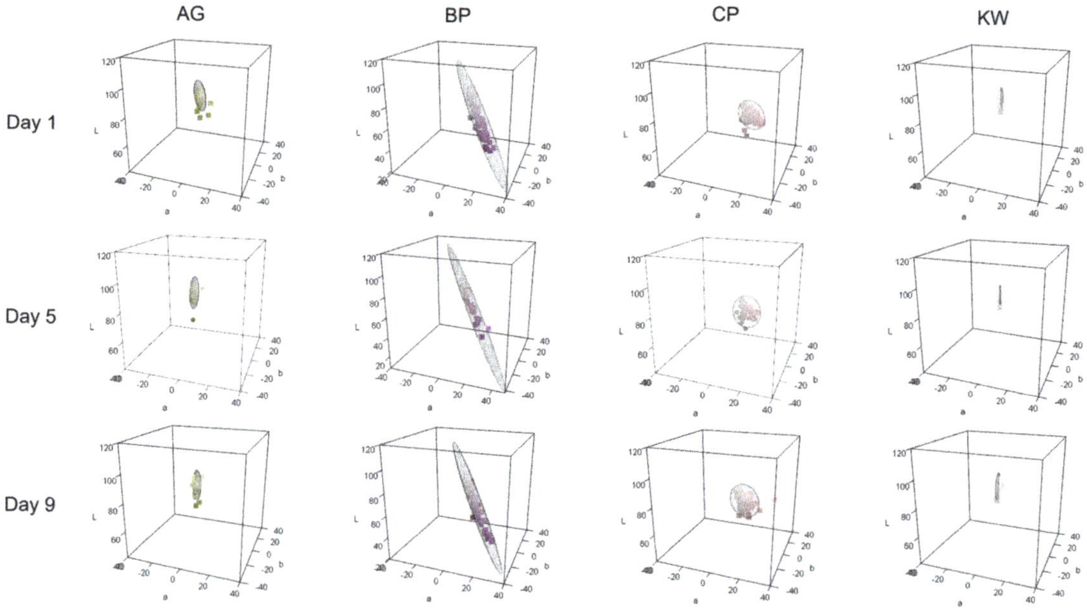

Figure 3. Three-dimensional plots indicating the color space (L*, a*, and b*) according to the measurement day (Day 1–9) for the vase life of four cultivars (AG, BP, CP, and KW) grown in soil cultivation.

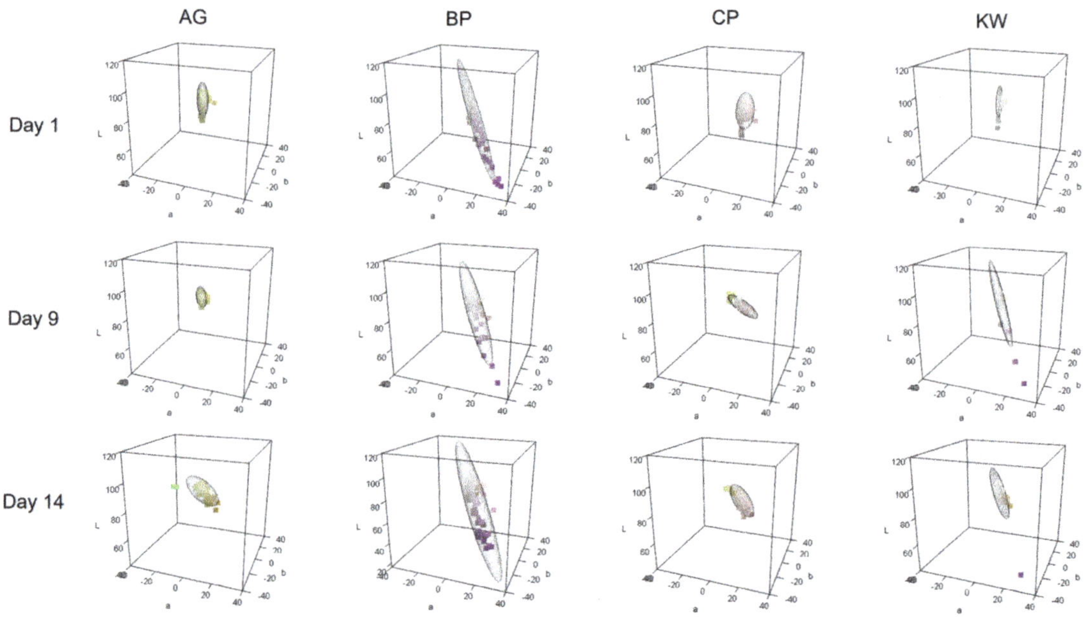

Figure 4. Three-dimensional plots indicating the color space (L*, a*, and b*) according to the measurement day (Day 1–14) for vase life of four cultivars (AG, BP, CP, and KW) grown in hydroponic cultivation.

3.2.1. Soil Cultivation Group

The average vase life of cut lisianthus grown in group S was approximately 10 days after flower cutting. Therefore, petal colorimetric data were collected from the day that flower cutting was performed (Day 1) to Days 5 and 9 (Figure 3). The change in the color space from Day 1 to Day 9 was not significant for any of the cultivars. The color space did not change significantly with changes in the measurement day of vase life, and as petal senescence progressed, the size of the ellipse did not change significantly. This means that, unlike in the H group, there was little variation between individuals, and the difference in color space between measurement days was not significant, making it difficult to distinguish based on petal color data between measurement days. While AG exhibited minimal variations in a* and b* values, the considerable variations in L* values resulted in an elongated oval shape aligned with the L* axis. Even as senescence progressed, the AG of the S group cultivars did not exhibit a significant increase in b* values. The ellipse of BP also did not change much with petal senescence, but the b* value increased and the a* value decreased, gradually weakening the characteristic purple color. The ellipse of CP has a larger b* value and a smaller a* value as it ages, meaning that the pink color of CP becomes less pink and the yellowish hue becomes more emphasized. KW did not show much variation in CIELAB values, forming a much smaller ellipse than the other varieties. However, as with the other varieties, there was a tendency for the b* value to increase on Day 9.

3.2.2. Hydroponic Cultivation Group

In the H group (Figure 4), the average vase life after flower cutting was approximately 15 days; therefore, the color space of the petals was examined on Days 1, 9, and 14. Because the H group had a relatively longer vase life than the S group, the petal color space differed significantly depending on the day of measurement. The color space on Day 1 for the three cultivars, except BP, was parallel to the L* axis. On Day 14, the color space tended to be parallel to the a* × b* plane, meaning that on Day 1, the difference in petal color was due to differences in brightness, while on Day 14, the variation in hue was greater than that in lightness.

Within the H group, AG did not exhibit differences in L*, a*, and b* values across the measurement days. However, the petal color variation increased on Day 14, resulting in the formation of a larger ellipse. On Day 14, the lime green color has appeared due to the relatively pale green and darker yellow color. BP demonstrated an increasing ellipse size as color senescence progressed. As senescence progresses, the variation in L, a, and b values increases significantly. The lightness of BP decreases and the red color becomes darker as senescence progresses. However, although the b* values of Day 9 and Day 14 are statistically different, it cannot be said that the b* value has decreased due to its large variability on Day 14. Also, the b* value is negative, indicating a blue color, but some samples on Day 14 have a positive value, showing that some individuals have a yellow color. Therefore, the tendency for b* values to increase with senescence can also be said to apply to BP in the H group. In the H group, both the CP and KW ellipses tended towards increasing b* values over time. Based on these aforementioned observations, particularly the distinct color space on Day 14 compared to the other measurement days, it was concluded that classification based on petal senescence is possible.

3.3. Classification Model for Predicting Vase Life Based on Biomass-Related Characteristics

Using the h2o package, we constructed a distributed random forest-based stacked ensemble model that could classify the day of vase life measurement based on petal colorimetric data (Figure 5).

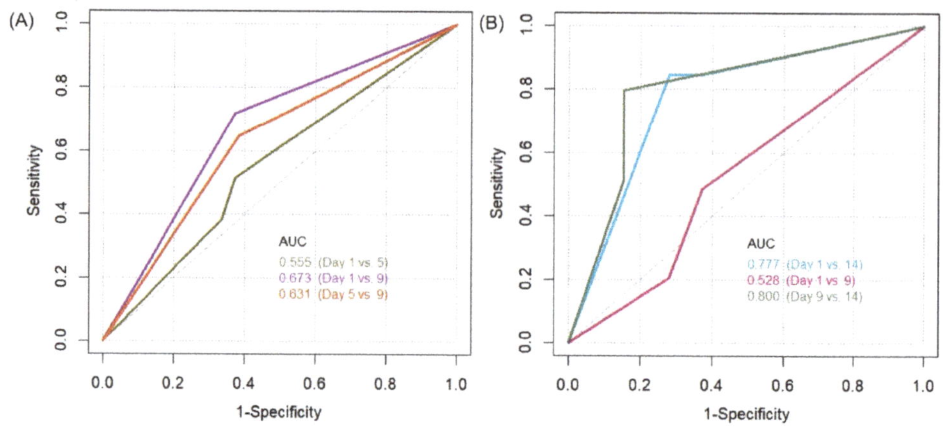

Figure 5. Multiclass receiver operating characteristic (ROC) curves and area under the ROC curve (AUC) of the top machine learning model selected by AutoML for classifying vase life based on petal color senescence of soil cultivation group (**A**) and hydroponic cultivation group (**B**). The AUC of each class is represented by the corresponding color in the graph.

The distinction between measurement days was clearer in the H group (Figure 5B) than in the S group (Figure 5A). In the H group, the distinction between Day 9 and Day 14 showed high performance with an AUC of 0.800. However, the distinction between Days 1 and 9 showed a relatively low performance, with an AUC of 0.528. This means that the difference of color space from flower cutting to Day 9 was not significant in the H group, and it can be seen that flower senescence progresses rapidly as it approaches Day 14. In other words, there is very little senescence progression from Day 1 to Day 9 in the H group, with a significant change in petal color at Day 14.

However, the S group showed a lower classification performance than the H group model, making it difficult to classify vase life based on the color space from petal senescence. The largest AUC values were observed between Days 1 and 9 (0.673), followed by Days 5 and 9 (0.631). However, it was difficult to distinguish between Day 1 and Day 5 (0.555). This indicated that in the S group, there was a sharp change in the color space due to senescence after Day 5.

4. Discussion

As shown in Figures 1 and 2, each cultivar has its own color pattern. However, they share the common characteristic of an increasing amount of yellow in the petals as senescence progresses. In other words, all cultivars showed an increasing trend in b* values as the senescence progressed. Specifically, regardless of S or H cultivation, on the day when the vase life reached its limit, the b* value increased significantly, differentiating it from other days. This is because, as senescence progresses, colorless or pale-yellow flavonols or flavones increase more than anthocyanins. This phenomenon is called copigmentation [42], and the intensity of the color change depends on the ratio of copigments to anthocyanins. In our previous study [29], we proposed the ratio between flavonol and anthocyanin content as a senescence index. As flower senescence progresses, the senescence index increases, indicating that vase life approaches its end. We are currently developing indirect and non-destructive phenomic methods using multispectral or hyperspectral cameras instead of directly measuring anthocyanins and other flavonoids using HPLC. Our next goal is to directly measure the amount of anthocyanins and flavonols in petals at different vase life stages using HPLC, and also to measure indices that are highly correlated with the content of each flavonoid using a hyperspectral camera. We plan to investigate the correlation between the calculated senescence index derived from HPLC results and spectral reflectance indices obtained from HSI.

As mentioned in the previous results, the BP of the H group showed the opposite trend to the common varieties, where the b* value decreased. However, as shown in Figure 2, this is likely due to the statistical error caused by the increased variability of the b* value of BP in the H group, which resulted in the average value being lower than those of other measurement days. In Figure 2, the b* value of the H group is much larger than that of the S group. In addition, the b* value of BP on Day 14 is much larger than that of other measurement days, with more individuals having positive values, and the value is as high as 20. In other words, BPs in the H group have the same tendency to increase the value of b* value as they senesce.

The best-performing model selected by AutoML was a stacked ensemble model using DRF and GBM as base models and GLM as the meta classifier. As shown in Figure 5A, comparison of Days 1 and 9 for S cultivation showed the best performance, with an AUC value of 0.673. Figure 5B shows the best performance, with an AUC value of 0.800, when comparing Days 9 and 14 of H cultivation. In H cultivation, it was difficult to distinguish vase life based on petal color until day 9; however, after day 9, this distinction became clear. Unlike H cultivation, S cultivation did not show the same level of performance in the differentiation of vase life stage, so we need to accumulate more data from our samples to refine our model parameters in the future.

This research is our third study investigating the vase life of cut lisianthus. Our first research has focused on improving the vase life of cut lisianthus through exogenous salicylic acid (SA) treatment and cultivation method (soil or hydroponic cultivation) [29]. That study revealed the effects of different cultivars (genotype) and growing conditions (environment), including SA treatment timing (vegetative or reproductive stage) and concentration (SA 0, 0.1, 0.3, or 0.5 mM), on improving vase life. It was confirmed that hydroponics and SA treatment at the reproductive stage increased the expected vase life after flower cutting, and the longest vase life was expected at a concentration of 0.5 mM SA for the KW cultivar among all cultivars and treatment groups [29]. Furthermore, we attempted to estimate senescence by observing and quantifying petal color changes in cut lisianthus. The automated machine learning models can improve the predictability of vase life based on the colorimetric (R^2 = 0.9551) and SPAD data (R^2 = 0.6966).

Our second study examined various biomass-related characteristics of the preharvest period under the same conditions as the first study and investigated their correlation with vase life, a postharvest characteristic [43]. The biomass-related characteristics are as follows: vegetative characteristics (stem diameter, number of stem nodes, stem length, number of stem bushes, flowering day, and SPAD value), reproductive characteristics (fresh weight, dry weight, weight difference, petal number, petal size), and leaf chemical components (nitrogen (N), phosphorus (P), potassium (K), magnesium (Mg), and calcium (Ca)). The second study showed that vase life has a high positive correlation with dry weight, SPAD, Mg content, and flowering day. As these characteristics directly or indirectly relate to photosynthesis, this result implies that increased photosynthetic efficiency significantly relates to improved vase life. The higher photosynthetic efficiency reflected in greater dry weight, which has a positive influence on vase life. The increase in dry weight serves as a source of carbohydrates that determine the longevity of cut flowers. In addition, we suggest that high tissue water content also contributes to extended vase life. Principle component analysis (PCA) revealed that BP cultivar's longer vase life is attributable to a combination of high carbohydrate content generated through photosynthesis and high tissue water content, both exceeding those observed in other cultivars [43]. We developed a machine learning model to estimate vase life based on preharvest biomass-related characteristics. According to the AutoML model and PCA result, dry weight was the strongest predictor of vase life. The R^2 value of this machine learning model based on biomass-related characteristics is 0.62.

In this study, our third, we attempted a classification analysis to distinguish the stages of vase life, rather than a regression analysis to estimate vase life as in the previous studies [29,43]. Specifically, regression models based on several biomass-related characteristics are less industrially valuable because each characteristic must be investigated during the lisianthus growing season. Therefore, to develop a model readily and immediately applicable in the floriculture industry, we attempted to create a model capable of discerning the stages of senescence directly from the cut flower state. As evident from our previous research findings, we aimed to target a single characteristic that is directly related to the vase life of cut flowers, rather than targeting multiple characteristics, so that it can be applied immediately in the industrial field. Therefore, we tried to develop a classification model based on petal color change using a colorimeter, as explored in our first study. With the increasing demand for cut flowers in the current flower market, it is crucial to conduct follow-up research to develop similar models for popular flower crops, such as roses, carnations, and chrysanthemums.

5. Conclusions

As senescence progresses after flower cutting, variations in the petal color space can be useful indicators for classifying the measurement day of vase life. The vase life classification model performed better for H-cultivated cut lisianthus than for S cultivation. Using this classification model, it is possible to distinguish the vase life stages, which are expected to be widely used throughout the floricultural industry in the future.

Author Contributions: Conceptualization, H.S.K. and S.H.; investigation, H.S.K.; data curation, H.S.K.; formal analysis, S.H.; software, S.H.; supervision, S.H.; writing—original draft, H.S.K. and S.H.; writing—review and editing, S.H. All authors have read and agreed to the published version of the manuscript.

Funding: This research received no external funding.

Data Availability Statement: Data are contained within the article.

Acknowledgments: This work was supported by the research grant of Kongju National University in 2024.

Conflicts of Interest: The authors declare no conflicts of interest.

References

1. Roh, M.S.; Lawson, R.H. Propagation and transplant production technology of new floral crops. In *Transplant Production Systems*, 1st ed.; Kurata, K., Kozai, T., Eds.; Springer: Dordrecht, The Netherlands, 1992; pp. 1–20.
2. Ahmad, H.; Rahu, S.K.; Mahbuba, S.; Jahan, M.R.; Uddin, A.J. Evaluation of Lisianthus (*Eustoma grandiflorum*) lines for commercial production in Bangladesh. *Int. J. Bus. Soc.* **2017**, *5*, 156–167.
3. Li, X.; Bu, F.; Wang, L.; Kim, C.; Xue, W.; Zhang, M.; Kawabata, L.; Zhang, Q.; Li, Y.; Zhang, Y. Optimization of CRISPR-Cas9 system in *Eustoma grandiflorum*. *iScience* **2024**, *27*, 109053. [CrossRef] [PubMed]
4. Lee, J.W.; Kim, S.D.; Lee, H.D.; Kim, J.H.; Yoon, T. *Research on Bellflower Plants Suppressed Short-Term Production through Growth Cultivation*, 1st ed.; Rural Development Administration: Jeonju, Republic of Korea, 2005.
5. Ohkawa, K.; Sasaki, E. *Eustoma* (Lisianthus)—Its past, present, and future. *Acta Hortic.* **1999**, *482*, 423–428. [CrossRef]
6. Halevy, A.H.; Kofranek, A.M. Evaluation of Lisianthus as a new flower crop. *Hortic. Sci.* **1984**, *19*, 845–847. [CrossRef]
7. Griesbach, R.J.; Bhat, R.N. Colchicine-induced polyploidy in *Eustoma grandiflorum*. *Hortic. Sci.* **1990**, *25*, 1284–1286. [CrossRef]
8. Sukpitak, C.; Seraypheap, K.; Muñoz, P.; Munné-Bosch, S. Influence of water deficit on the longevity of ethylene-sensitive and ethylene-insensitive flowers. *Environ. Exp. Bot.* **2024**, *219*, 105647. [CrossRef]
9. Aktas, H.; Akkaya, G.G.; Cagdas, E.A.; Haspolat, G.; Ellialtioglu, S.S. Micropropagation of lisianthus (*Eustoma grandiflorum* Raf. Shinn). *Hortic. Sci.* **2023**, *40*, 86–97.
10. Liang, Y.; Zhao, P.; Liu, B.; Sun, D.; Ruan, J.; Jin, C.; Fang, H.; Wang, Q.; Sun, F.; Bi, B.; et al. Genetic mechanisms of petal morphogenesis in *Eustoma grandiflorum*. *Sci. Hortic.* **2024**, *324*, 112558. [CrossRef]
11. Fang, F.; Oliva, M.; Ovadia, R.; Bar, E.; Nissim-Levi, A.; Kumar, V.; Wang, R.; Neeman, A.; Zaccai, M.; Lewinsohn, E.; et al. Increased substrate availability reveals the potential of scentless lisianthus flowers in producing fragrant benzenoid-phenylpropanoids. *Physiol. Plant.* **2021**, *172*, 19–28. [CrossRef]
12. Roni, M.Z.K.; Islam, M.S.; Shimasaki, K. A timeline for *Eustoma grandiflorum* seedling production based on an in vitro germination protocol. *Plant Omics* **2017**, *10*, 232–236. [CrossRef]
13. Höhn, D.; Peil, R.M.N.; Marchi, P.M.; Grolli, P.R.; Trentin, R.; Shaun, W.S. Base temperature estimates for lisianthus cultivars grown in different planting seasons. *Pesqui. Agropecu. Bras.* **2023**, *58*, e03447. [CrossRef]
14. Jones, R.B.; Hill, M. The effect of germicides on the longevity of cut flowers. *J. Am. Soc. Hortic. Sci.* **1993**, *118*, 350–354. [CrossRef]

15. Ezhilmathi, K.; Singh, V.P.; Arora, A.; Sairam, R.K. Effect of 5-sulfosalicylic acid on antioxidant activity in relation to vase life of gladiolus cut flowers. *Plant Growth Regul.* **2007**, *51*, 99–108. [CrossRef]
16. Kazemi, M.; Abdossi, V.; Kalateh Jari, S.; Ladan Moghadam, A.R. Effect of pre-and postharvest salicylic acid treatment on physio-chemical attributes in relation to the vase life of cut rose flowers. *J. Hortic. Sci. Biotechnol.* **2018**, *93*, 81–90. [CrossRef]
17. Alaey, M.; Babalar, M.; Naderi, R.; Kafi, M. Effect of pre-and postharvest salicylic acid treatment on physio-chemical attributes in relation to vase-life of rose cut flowers. *Postharvest Biol. Technol.* **2011**, *61*, 91–94. [CrossRef]
18. Hurr, B.M.; Huber, D.J.; Vallejos, C.E. Features of programmed cell death precede water soaking development in ethylene-treated immature cucumber fruit. *Postharvest Biol. Technol.* **2010**, *58*, 13–20. [CrossRef]
19. Iqbal, N.; Khan, N.A.; Ferrante, A.; Trivellini, A.; Francini, A.; Khan, M.I.R. Ethylene role in plant growth, development and senescence: Interaction with other phytohormones. *Front. Plant Sci.* **2017**, *8*, 475. [CrossRef]
20. Liu, J.; Zhang, Z.; Li, H.; Lin, X.; Lin, S.; Joyce, D.C.; He, S. Alleviation of effects of exogenous ethylene on cut 'Master' carnation flowers with nano-silver and silver thiosulfate. *Postharvest Biol. Technol.* **2018**, *143*, 86–91. [CrossRef]
21. Darvish, M.; Shirzad, H.; Asghari, M.; Noruzi, P.; Alirezalu, A.; Pateiro, M.; Takshe, A.; Lorenzo, J.M. 24-Epibrasinolide modulates the vase life of lisianthus cut flowers by modulating ACC oxidase enzyme activity and physiological responses. *Plants* **2021**, *10*, 995. [CrossRef]
22. Yadeta, K.A.J.; Thomma, B.P. The xylem as battleground for plant hosts and vascular wilt pathogens. *Front. Plant Sci.* **2013**, *4*, 97. [CrossRef]
23. Gómez-Merino, F.C.; Ramírez-Martínez, M.; Castillo-González, A.M.; Trejo-Téllez, L.I. Lanthanum prolongs vase life of cut tulip flowers by increasing water consumption and concentrations of sugars, proteins and chlorophylls. *Sci. Rep.* **2020**, *10*, e4209. [CrossRef] [PubMed]
24. Da Costa, L.C.; De Araujo, F.F.; Ribeiro, W.S.; De Sousa Santos, M.N.; Finger, F.L. Postharvest physiology of cut flowers. *Ornam. Hortic.* **2021**, *27*, 374–385. [CrossRef]
25. Liu, Z.; Luo, Y.; Liao, W. Postharvest physiology of fresh-cut flowers. In *Oxygen, Nitrogen and Sulfur Species in Post-Harvest Physiology of Horticultural Crops*, 1st ed.; Ziogas, V., Corpas, F.J., Eds.; Academic Press: Cambridge, MA, USA, 2024; pp. 23–42.
26. Kaviani, B.; Zamiraee, F.; Zanjani, S.B.; Tarang, A.; Torkashvand, A.M. In vitro flowering and micropropagation of lisianthus (*Eustoma grandiflorum*) in response to plant growth regulators (NAA and BA). *Acta Sci. Pol. Hortorum Cultus* **2014**, *13*, 145–155.
27. Zhou, X.; Li, C.; Liu, L.; Zhao, J.; Zhang, J.; Cai, Z.; Huang, X. Control of Fusarium wilt of lisianthus by reassembling the microbial community in infested soil through reductive soil disinfestation. *Microbiol. Res.* **2019**, *220*, 1–11. [CrossRef] [PubMed]
28. Wang, X.; Chen, Z.; Sui, N. Sensitivity and responses of chloroplasts to salt stress in plants. *Front. Plant Sci.* **2024**, *15*, 1374086. [CrossRef]
29. Kwon, H.S.; Leporini, C.; Kim, S.; Heo, S. Prolonged vase life by salicylic acid treatment and prediction of vase life using petal color senescence of cut lisianthus. *Postharvest Biol. Technol.* **2024**, *209*, 112726. [CrossRef]
30. Teerarak, M.; Pilasombut, K.; Laosinwattana, C. Peppermint essential oil enhances the vase life of Dendrobium orchids. *Heliyon* **2024**, *10*, e31636. [CrossRef]
31. Kumarihami, H.M.P.C.; Wijerathna, Y.M.A.M.; Beneragama, C.K. Extending the postharvest life of cut flowers with floral preservatives: A mini review. In Proceedings of the National Symposium on Floriculture Research, Peradeniya, Sri Lanka, 31 October 2017; pp. 52–60.
32. Kim, H.S.; Chung, Y.C. Measurement of anthocyanin accumulations in multiple seedling plants using hyperspectral imaging technology. *Korean J. Opt. Photon.* **2021**, *32*, 215–219.
33. Gabrielli, M.; Lançon-Verdier, V.; Picouet, P.; Maury, C. Hyperspectral imaging to characterize table grapes. *Chemosensors* **2021**, *9*, 71. [CrossRef]
34. Choudhury, A.K.R. Colour measurement instruments. In *Principles of Colour and Appearance Measurement*; Woodhead Publishing: Cambridge, UK, 2014; pp. 221–269.
35. Sanmartín, N.; Pastor, V.; Pastor-Fernández, J.; Flors, V.; Pozo, M.J.; Sánchez-Bel, P. Role and mechanisms of callose priming in mycorrhiza-induced resistance. *J. Exp. Bot.* **2020**, *71*, 2769–2781. [CrossRef]
36. Braun, G.J.; Fairchild, M.D.; Ebner, F. Color gamut mapping in a hue-linearized CIELAB color space. In *Color and Imaging Conference*; Society of Imaging Science and Technology: Springfield, VA, USA, 1998; Volume 6, pp. 163–168.
37. Uddin, A.J.; Hashimoto, F.; Miwa, T.; Ohbo, K.; Sakata, Y. Seasonal variation in pigmentation and anthocyanidin phenetics in commercial *Eustoma* flowers. *Sci. Hortic.* **2004**, *100*, 103–115. [CrossRef]
38. Cabahug, R.A.M.; Nam, S.Y.; Lim, K.B.; Jeon, J.K.; Hwang, Y.J. Propagation techniques for ornamental succulents. *Flower Res. J.* **2018**, *26*, 90–101. [CrossRef]
39. Thi, M.K.H.T.; Lim, K.B.; Hwang, Y.J. Phenotype and ploidy analysis of the colchicine-induced M_1 generation of *Echeveria* species. *Hortic. Sci. Technol.* **2020**, *38*, 522–537.
40. R Core Team. R: A Language and Environment for Statistical Computing. R Foundation for Statistical Computing. 2023. Available online: https://www.R-project.org/ (accessed on 10 July 2024).
41. Fryda, T.; LeDell, E.; Gill, N.; Aiello, S.; Fu, A.; Candel, A.; Click, C.; Kraljevic, T.; Nykodym, T.; Aboyoun, P.; et al. H_2O: R Interface for the 'H_2O' SCALABLE Machine Learning Platform. R Package Version 3.40.0.4. 2023. Available online: https://CRAN.R-project.org/package=h2o (accessed on 10 July 2024).

42. Qin, Q.; Tatsuzawa, F.; Nakane, T.; Kaidzuka, T.; Iwashina, T.; Mizuno, T. Anthocyanins and flavonols from the flowers o *Ranunculus* cultivars (Ranunculaceae) and their color expression. *Hortic. J.* **2024**, *93*, 114–125. [CrossRef]
43. Kwon, H.S.; Heo, S. Predicting vase life of cut lisianthus based on biomass-related characteristics using AutoML. *Agriculture* **2024** *14*, 1543. [CrossRef]

horticulturae

MDPI

Article

Characterization of *AGAMOUS* Ortholog and Promoter from the *Ilex verticillata* (Aquifoliaceae)

Jiayi Li †, Yalan Su †, Xiangjian Chen and Zhixiong Liu *

College of Horticulture and Gardening, Yangtze University, Jingzhou 434025, China;
2021710820@yangtzeu.edu.cn (J.L.); 2022710854@yangtzeu.edu.cn (Y.S.); 2023730097@yangtzeu.edu.cn (X.C.)
* Correspondence: zxliu@yangtzeu.edu.cn
† These authors contributed equally to this work.

Abstract: *Arabidopsis AGAMOUS (AG)* plays a crucial role in specifying stamen and carpel identities, floral meristem identity determination, and repression of the A-function. *Ilex verticillata* (Aquifoliaceae) is a dioecious shrub, whereby an individual plant has either male or female flowers with vestigial organs of the opposite sex. The molecular mechanism of male and female organ development in *I. verticillata* remains unknown. In order to identify the possible roles of *AG*-like genes in regulating floral development in *I. verticillata, AG* ortholog (*IlveAG*) and its promoter (*pIlveAG)* from the male and female flowers of *I. verticillata* were separately isolated. *IlveAG* is highly expressed in stamens, pistils, and sepals of male and female flowers. Moreover, obvious GUS staining was observed in the inflorescence and sepals, stamens, and pistils of mature flowers in *pIlveAG::GUS Arabidopsis*. The 35S::*IlveAG Arabidopsis* showed obviously early flowering. Moreover, *IlveAG* could substitute for endogenous *AG* to rescue the stamen and pistil in the *Arabidopsis ag-1* mutant. In addition, expression of *IlveAG* can inhibit the development of sepals and petals (two outer whorls of floral organs) in wild-type and *ag-1 Arabidopsis*. Our findings suggest that *IlveAG* has a conservative C-function and plays key roles in determination of reproductive floral organs (stamen and carpel) identity and meristem determinacy. Our results provide more details to understand the role of *AG* orthologs in the development of male and female flowers in woody plants.

Keywords: *AGAMOUS*; *Ilex verticillata*; flower development

1. Introduction

Ilex verticillata (L.) A. Gray (Aquifoliaceae) is a deciduous shrub originating from North America and is widely cultivated in many countries in Europe and America. Moreover, Winterberry Holly (*I. verticillata*) was introduced and widely cultivated in different areas of China for ornamental red winter fruits. *I. verticillata* is also a very popular ornamental cut flower or potted plant at Chinese Lunar New Year flower markets. In addition, *I. verticillata* is also an interesting dioecious shrub, whereby an individual plant has either male or female flowers with vestigial organs of the opposite sex. The male flower possesses a vestigial pistil (pistillode), and the female flower possesses vestigial stamens (staminodes) (Figure 1), which make it an excellent model for exploring male and female flower development during origin and evolution of dioecious angiosperm. Moreover, the flowering time is synchronized between male and female trees, resulting in a low fruit setting rate. Improving the pollination and fruit setting rate of *I. verticillata* requires a better understanding of the molecular mechanisms of the male and female flower development. Previous studies indicated that different floral homeotic MADS transcription factors (TFs) work together to determine the development of different types of floral organs during flower development [1,2].

In *Arabidopsis*, the C-class MADS-box TF AGAMOUS (AG) plays central roles in determination of reproductive floral organ (stamen and carpel) identity and meristem determinacy [3]. In addition, *AG* is closely related to the process of sex differentiation, flowering, and fruiting. Previous studies showed that AG orthologs from other rosid species, *PrseAG* from *Prunus lannesiana* [4], *PPERAG* from *P. persica* [5], *KjAG* from *Kerria japonica* [6], and *EjAG* from *Eriobotrya japonica* [7], were mainly expressed in stamen and carpel, and also regulated stamen and carpel development, which showed the conservative expression pattern and function of *AG* orthologs among rosid species.

Figure 1. Female and male flower of *Ilex verticillata* (L.) A. Gray. (**A**,**B**) Female flower with 7 staminodes; (**C**,**D**) Male flower with a pistillode. Petal (pet), staminode (stn), pistil (pis), stamen (sta), pistillode (pil). Scale bar = 1 mm.

In this study, we isolated an *AG* ortholog (*IlveAG*) and its promoter (*pIlveAG*) from both the male and female flowers of *I. verticillata*. Moreover, we analyzed the function of *IlveAG* and *pIlveAG* by detecting the expression pattern of the gene and the activity of the promoter, and combining the ectopic expression of the gene and the promoter in *Arabidopsis*. The flowers of male and female trees of *I. verticillata* consist of four whorls of floral organs (5–9 sepals in whorl 1, 5–9 petals in whorl 2, 5–9 stamens (staminodes in female flowers) in whorl 3, and 1 pistil (pistillodes in male flowers) in whorl 4 (Figure 1). Our results provide more details for understanding the function of the *AG* ortholog in male and female flowers of deciduous woody plants.

2. Materials and Methods

2.1. Plant Material and Growth Conditions

Male and female floral buds at sequential development stages were separately sampled from male/female trees growing under natural conditions in the campus of Yangtze University (Jingzhou, Hubei, China). Moreover, male and female floral buds were divided in half, and half of them were immediately put into liquid nitrogen and then stored at −80 °C. The remains were fixed in FAA (38% formaldehyde: acetic: 70% ethanol = 1:1:18, by volume). The root, stem, leaf, sepal, petal, stamen, and vestigial pistil of male trees, as well as the root, stem, leaf, sepal, petal, vestigial stamen, pistil, and young fruit of female trees were separately dissected and immediately frozen in liquid nitrogen for RNA isolation. The *ag-1* mutant *Arabidopsis* (CS25) seeds were obtained from the ABRC (*Arabidopsis* Biological Resource Center) at Ohio State University, Columbus, OH, USA.

2.2. Isolation of IlveAG and IlveAG Promoter (pIlveAG) from I. verticillata

Total RNA was separately extracted from male and female floral buds using the EASYspin Total RNA Kit (Aidlab, Beijing, China) referring to the manufacturer's protocol. The *AG* ortholog *IlveAG* was separately isolated from male and female floral buds by 3′-full RACE Core Set Ver. 2.0 kit (TaKaRa, Shiga, Japan) with forward primer GSPIVAG3 according to manufacturer's protocol (Table S1). The forward primer was designed based on the transcript (transcript_3468) of DEG (differentially expressed gene) (Ilumina and PacBio sequencing in Biomarker Technologies) (Biomarker Technologies, Beijing, China). Phylogenetic analysis of *IlveAG* was referenced to Li et al. [8]. All *AG* homolog proteins from other species were selected for phylogenetic trees obtained from NCBI Genbank (Table S2).

I. verticillata genomic DNA was extracted from leaves using the CTAB Plant Genomic DNA Rapid Extraction Kit (Aidlab, Beijing, China) following the manufacturer's instructions. The *IlveAG* promoter was cloned from *I. verticillata* genomic DNA with the Genome Walking Kit (TaKaRa, Shiga, Japan), referring to the manufacturer's guidelines.

The forward primers GSPilveAG-1R, GSPilveAG-2R, and GSPilveAG-3R (Table S1) were selected for amplifying the first target sequence, and GSPilveAG-4-1R, GSPilveAG-4-2R, and GSPilveAG-3-3R (Table S1) were selected for amplifying the second target sequence. The *IlveAG* gene and its promoter were sequenced by using Sanger sequencing in Sangon Biotech (Sangon, Shanghai, China). The cis-acting elements lying at the *IlveAG* promoter (*pIlveAG*) were searched in the PLACE database [9].

2.3. Characterization of pIlveAG Activity from the Promoter in Transgenic Arabidopsis

The *pIlveAG* fragment was introduced into the pCAMBIA1300 vector with *Xba* I (TaKaRa, Shiga, Japan) and *Sac* I (TaKaRa, Shiga, Japan) by using the ClonExpress® Ultra One Step Cloning Kit (Vazyme, Nanjing, China), referring to the manufacturer's instructions. The *pIlveAG* was cloned with the primer pair pIlveAG-F and pIlveAG-R (Table S1). Then, the construct was transformed into *Col-0 A. thaliana* using the floral-dip method described by Clough and Bent [10]. Transgenic *Arabidopsis* seedings were obtained and cultivated for histochemical GUS staining according to Liu et al. [11]. The procedure and observation of GUS staining referred to You et al. [12].

2.4. Cytomorphological Examination and Expression Analysis of IlveAG

The male and female floral buds samples of *I. verticillata* fixed in FAA were separately dehydrated in a graded ethanol series, cleared with a xylene series, infiltrated with molten paraffin, and then embedded into paraffin block, which was serially sectioned with a Leica RM2235 rotary microtome, and then these sections were stained with safranin-fast green using the method suggested by Liu et al. [13]. Each section was microscopically examined using a CAIKON RCK-40C microscope and photomicrographs were taken.

Total RNA of floral buds at sequential developmental stages were separately extracted from male/female trees with EASYspin Total RNA Kit (Aidlab, Beijing, China) according to the manufacturer's instructions. The first-stand cDNA was synthesized for qRT-PCR with HiScript® II Q RT SuperMix for qPCR kit (Vazyme, Nanjing, China) according to the manufacturer's instructions. *IlveAG* expression was separately measured in root, stem, leaf, sepal, petal, stamen, and vestigial pistils of male trees, and in root, stem, leaf, sepal, petal, vestigial stamen, pistil, and young fruit of female trees by qRT-PCR, referring to Liu et al. [11], but with primer pair qIlveAGF and qIlveAGR (Table S1). In addition, *IlveAG* expression was also separately detected in male and female floral buds at sequential development stages. An amplification fragment of the *I. verticillata* actin gene was selected as the internal control with primer pair qIlveactinF and qIlveactinR. The qRT-PCR was performed with three biological replicates and relative expression levels were detected, referring to Liu et al. [11], but with 30 s annealing at 57 °C.

2.5. Ectopic Expression Analysis of IlveAG in Arabidopsis ag-1 Mutant

Full-length *IlveAG* cDNA was cloned into the pBI121 vector with *Xba* I and *Sac* I using the ClonExpress® Ultra One Step Cloning Kit (Vazyme, Nanjing, China), referring to the manufacturer's instructions, but with primer pair TIlveAGF2 and TIlveAGR (Table S1). The *35S::IlveAG* construct was introduced into the heterozygous *AG/ag-1 Arabidopsis* with the above method. Transgenic *Arabidopsis* seeds were screened and seedlings were cultivated in a growth chamber according to Li et al. [14]. Wild-type, heterozygous *AG/ag-1*, and homozygous *ag-1* transgenic *Arabidopsis* lines were confirmed by the dCAPS finder program described by Neff et al. [15]. The transgenic *Arabidopsis* phenotypes were analyzed after flowering.

3. Results

3.1. Characterization of IlveAG and IlveAG Promoter (pIlveAG) from I. verticillata

The 1088 bp *IlveAG* cDNA has a 726 bp ORF (Open Reading Frame), which encodes 241aa (amino acids) (Genbank accession number: PP415536). Phylogenetic tree analysis grouped IlveAG into core eudicots with euAG lineage (Figure 2), and the gene was named as *IlveAG* (*Ilex verticillata AGAMOUS*).

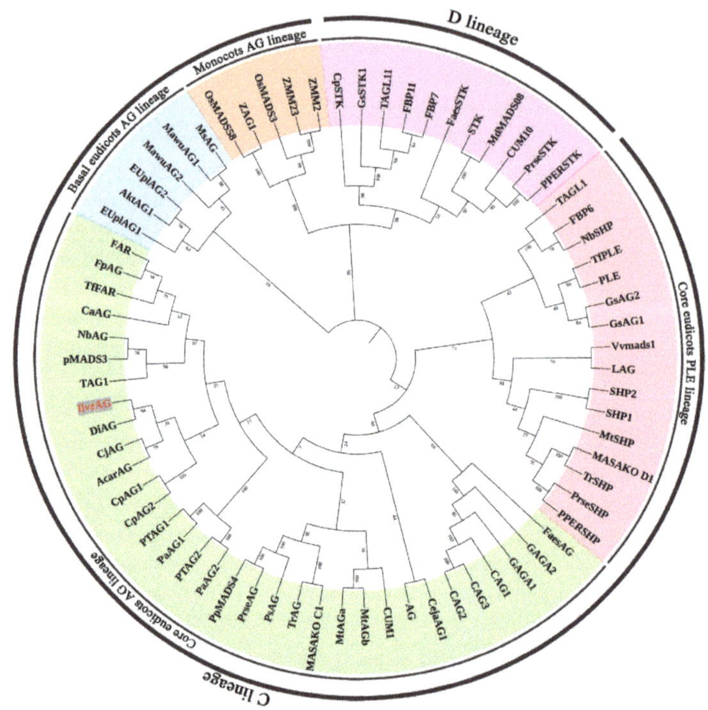

Figure 2. Phylogenetic tree of IlveAG and AG-like transcription factors from other species of angiosperms.

A 1669 bp *IlveAG* promoter (*pIlveAG*) (−1297/+372) was cloned from *I. verticillata* (Figure S1). The *pIlveAG* contains eight POLLEN1LELAT52-box and six GTGANTG10-box [16,17], which are usually located at the promoter of the genes involved in stamen development and suggest that *IlveAG* may be involved in regulating stamen development in *I. verticillata*. The *pIlveAG* also contains two CArG-box for floral homeotic MADS-box TFs [18] and one CCAAT-box for CONSTANS TF binding to promote flowering [19], as well as two AACAAA/TTTGTT motifs for floral homeotic APETALA2 TF binding [20]. All these data suggest that *pIlveAG* may drive the *IlveAG* gene to promote flowering and regulate floral organ development. In addition, gibberellin-responsive elements (PYRIMIDINEBOXOSRAMY1A-box) [21], abscisic-acid-responsive elements (MYB1AT) [22], drought-stress-responsive elements (MYBCORE) [23], and low-temperature-stress-responsive elements (MYCCONSENSUSAT) [24] were found in *pIlveAG*. These data suggest that hormone and stress signaling pathways may also directly induce *IlveAG* expression.

3.2. Expression Activity Analysis of pIlveAG in Transgenic Arabidopsis

A *GUS* reporter gene driven by *pIlveAG* was assayed in transgenic *Arabidopsis*. GUS staining was examined in the T_1 generation of *pIlveAG::GUS* independent transgenic lines (Figure 3). Intense GUS staining was obviously observed in the inflorescence and sepals, stamens (anther and filament), and pistils (stigma and style) of mature flowers, but was absent in petals. In addition, GUS reporter gene driven by *pIlveAG* was activated in sepals, filaments, and pistils of floral buds at stage 12, but was absent in petals and anthers. These data suggest that the *pIlveAG* may drive *IlveAG* to regulate inflorescence and floral organ development in *I. verticillata*.

Figure 3. Histochemical GUS staining of *pIlveAG* promoter in T_1 transgenic *Arabidopsis thaliana*. (**A**) Inflorescence of wild-type *Arabidopsis*; (**B**) flower of wild-type *Arabidopsis*; (**C**) floral bud at stage 12 in wild-type *Arabidopsis*; (**D**) inflorescence of *pIlveAG::GUS* transgenic *Arabidopsis*; (**E**) flower of *pIlveAG::GUS* transgenic *Arabidopsis*; (**F**) floral bud at stage 12 in *pIlveAG::GUS* transgenic *Arabidopsis*. Sepal (sep), petal (pet), anther (ant), filament (fil), stigma (sti), style (sty), ovary (ova); scale bars: (**A**,**B**,**D**,**E**) 1 mm, (**C**,**F**) 500 μm.

3.3. Expression Analysis of *IlveAG* in *I. verticillata*

IlveAG was obviously expressed in all floral organs and young fruits of the *I. verticillata*, but was absent in roots, stems, and leaves (Figure 4C). In the male/female trees, the expression level of the *IlveAG* in stamens was significantly higher than that in sepals, petals, pistils, and young fruit (*LSD*, $p < 0.01$). The expression level in petals was lowest among the four whorl floral organs (*LSD*, $p < 0.01$). In addition, obvious expression of *IlveAG* was also detected during the male/female floral bud differentiation. Furthermore, the expression level of *IlveAG* reached the peak at pollen maturation of male floral buds, which was significantly higher than that at other developmental stages (*LSD*, $p < 0.01$) (Figure 5A,B). In the female flower buds, the expression of *IlveAG* decreased in the stage of linear tetrad of megasporocyte (Figure 5A,C).

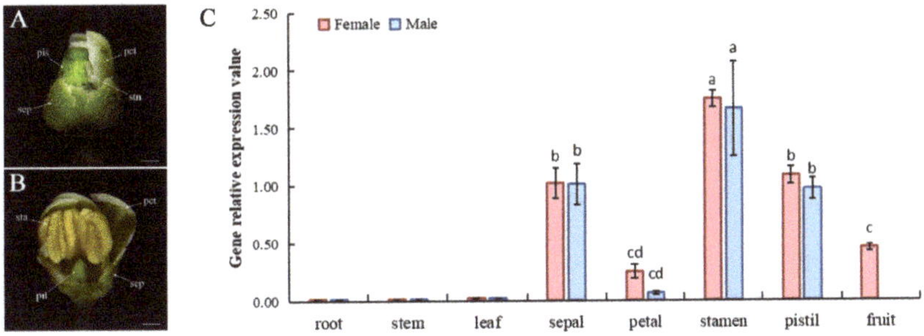

Figure 4. The expression of *IlveAG* in male and female trees of *I. verticillata*. (**A**,**B**) Flower bud structure of male and female trees in *I. verticillata*; (**C**) *IlveAG* expression in the root, stem, leaf, sepal, petal, stamen, pistil, and fruit were measured by qRT-PCR with *IlveAG* actin as the internal control. Scale bar: (**A**,**B**) 500 μm; Different letters indicate a significant difference ($p < 0.01$, LSD).

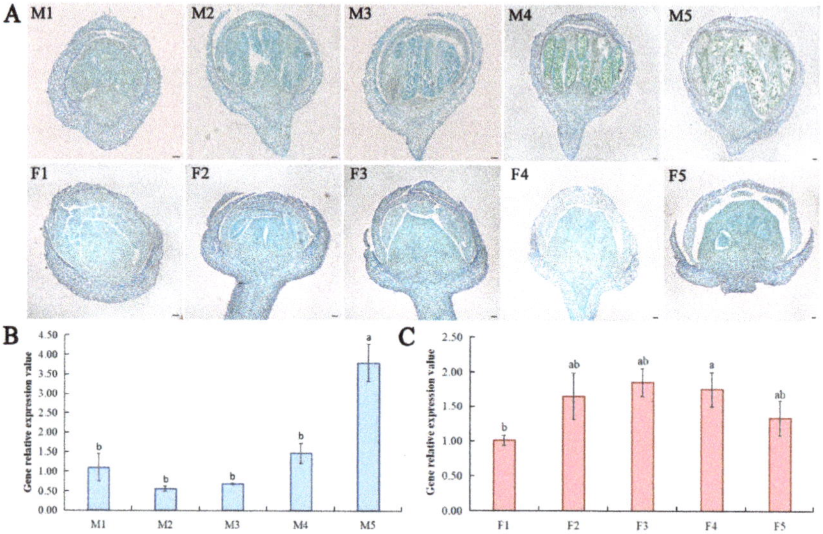

Figure 5. The expression of *IlveAG* in male and female floral buds of *I. verticillata* at different developmental stages. (**A**) Cytomorphological section of floral male buds at different development stages, M1: microspore mother cell formation; M2: microspore mother cell meiosis; M3: microspore tetrad formation; M4: mononuclear microspore in nucleus located aside stage; M5: full-maturity flower buds with pollen maturation. Cytomorphological section of female buds at sequential development stages, F1: microspore mother cell formation and pistil primodium development; F2: ovule appearance; F3: meiosis period of megasporocyte; F4: linear tetrad period of megasporocyte; F5: embryo sac maturation. (**B**) The expression of *IlveAG* in different stages of male flower buds; (**C**) the expression of *IlveAG* in different developmental stages of female flower buds; scale bar: (**A**) 100 µm; different lowercase letters show significant differences ($p < 0.01$ by LSD).

3.4. Ectopic Expression of IlveAG in ag-1 Arabidopsis Mutant

To evaluate whether *IlveAG* could complete the loss-of-function *AG* phenotype in *ag-1* mutant *Arabidopsis, 35S::IlveAG* was transformed into heterozygote *AG/ag-1 Arabidopsis* to create complementation lines through *agrobacterium* mediation. Independent transgenic *Arabidopsis* lines were identified by dCAPS genotyping. Moreover, twelve *35S::IlveAG* lines in the wild-type background and eleven independent *35S::IlveAG* lines in the homozygous mutant *ag-1* background were obtained. In addition, phenotypes of transgenic *Arabidopsis* lines in the wild-type and homozygous *ag-1* mutant backgrounds were separately assayed to evaluate whether *IlveAG* could mimic endogenous *AG* in specifying stamen and carpel identity.

Among twelve *35S::IlveAG* transgenic *Arabidopis* lines in the wild-type background, all lines were observed obviously flowering early. Seven (58.33%) lines had floral organ phenotype changes (Figure 6C–I); one (8.33%) showed its sepal converted into a carpeloid organ (Figure 6C); one (8.33%) showed a small sepal, petal, and stamen (Figure 6D); and five (41.67%) showed decreased sepal numbers (Figure 6E–I), which suggested that *IlveAG* conserved C-function activity and inhibited the development of the outer two whorls of floral organs (sepal and petal).

Figure 6. Phenotypes of *35S::IlveAG* transgenic *Arabidopsis* in the wild-type background. (**A**) Wild-type *Arabidopsis* with normal raceme; (**B**) normal flower of wild-type *Arabidopsis thaliana* (1st whorl with 4 sepals, 2nd whorl with 4 petals, 3rd whorl with 6 stamens, 4th whorl with 1 pistil); (**C**) *35S::IlveAG* transgenic *Arabidopsis* flower with one sepal transformed into a carpeloid organ and the number of sepals, petals, and stamens reduced; (**D**) *35S::IlveAG* transgenic *Arabidopsis* flower with smaller sepals, petals, and stamens; (**E**) *35S::IlveAG* transgenic *Arabidopsis* flower with smaller sepals and petals; (**F**) *35S::IlveAG* transgenic *Arabidopsis* flower with fewer sepals, petals, and stamens; (**G**) *35S::IlveAG* transgenic *Arabidopsis* flower with small sepals; (**H**) *35S::IlveAG* transgenic *Arabidopsis* flower with short stamens and fewer sepals; (**I**) *35S::IlveAG* transgenic *Arabidopsis* flower with small stamens. Sepal (sep), petal (pet), anther (ant), pistil (pis), stigma (sti), ovule (ovu). Scale bars: 1 mm.

Among eleven *35S::IlveAG* transgenic *Arabidopis* lines in the homozygous *ag-1* mutant backgrounds, one (9.10%) produced flowers only with pistil and carpeloid sepals (Figure 7C), one (9.10%) produced flowers with normal pistils, partial stamens, and one sepal transformed into pistil (Figure 7D). Two (18.18%) lines completely restored stamens and pistils (Figure 7E,F), and one (9.10%) produced partial filaments and carpels (Figure 7G). The infinite development pattern of flowers was terminated. Two (18.18%) lines displayed a small number of stamens within the sepal and petal nested pattern (Figure 7H,I). The remaining four (36.36%) lines were similar to the homozygous *ag-1* mutant *Arabidopsis* flowers (Figure 7B).

Figure 7. Phenotypes of *35S::IlveAG* homozygous transgenic *Arabidopsis ag-1* mutants. (**A**) *Arabidopsis ag-1* mutant raceme; (**B**) homozygous *Arabidopsis ag-1* mutant flower (sepals in the 1st whorl, petals in the 2nd and 3rd whorls, and reiterations of this pattern in interior whorls); (**C**) *35S::IlveAG* transgenic *ag-1 Arabidopsis* flower with pistil and carpeloid sepal; (**D**) *35S::IlveAG* transgenic *ag-1 Arabidopsis* flower with normal pistil, partial stamen, and one sepal transformed into pistil; (**E**) *35S::IlveAG* transgenic *ag-1 Arabidopsis* flower with restored pistils and stamens; (**F**) *35S::IlveAG* transgenic *ag-1 Arabidopsis* flower recovered four whorls of floral organs; (**G**) *35S::IlveAG* transgenic *ag-1 Arabidopsis* flower with complementation of filaments and carpels; (**H**) *35S::IlveAG* transgenic *ag-1 Arabidopsis* flower with the appearance of filaments; (**I**) *35S::IlveAG* transgenic *ag-1 Arabidopsis* flower with the appearance of filaments and partial carpels. Sepal (sep), petal (pet), anther (ant), pistil (pis), stigma (sti), ovule (ovu), filament (fil), carpel (car). Scale bars: 1 mm.

4. Discussion

I. verticillata is a dioecious shrub, whereby an individual plant has either male or female flowers with vestigial organs of the opposite sex. The male flower contains pistillodes, while the female flower contains staminodes, which make it an excellent model for exploring male and female flower development during origin and evolution of dioecious angiosperm. In *Arabidopsis*, *AG* plays key roles in controlling reproductive floral organ (stamen and carpel) identity and determining floral meristem identity [3]. Moreover, *AG* and *APETALA2* (*AP2*) represent C- and A-class genes that work in an antagonistic fashion to specify reproductive organs (stamen and pistil) and perianth (sepal and petal), respectively [25]. AP2 binds at the *AG* second intron and directly restricts *AG* expression extending to the outer two whorls [20]. The expression of *AG* begins at stage 3 of flower development, when stamen and pistil primordia arise, and its expression persists throughout the development of the stamen and pistil [26]. In *I. verticillata, IlveAG* was obviously expressed in all floral organs and was highest in stamens (staminodes). Functional analysis indicated that *IlveAG* could mimic endogenous *AG* in determining stamen and carpel identity in *ag-1* mutant *Arabidopsis*. In rosid species, *AG* orthologs, such as *TrAG* from *T. rupestris* [27], *MASAKO C1* from *Rosa rugosa* [28], and *EjAG* from *Eriobotrya japonica* [7], were mainly expressed in stamens and pistils. Moreover, ectopic expression of each gene showed similar results with *IlveAG* and exhibited a conserved function in specifying reproductive (stamen and pistil) organ identity.

In asterids species, *PMADS3* from *Petunia hybrida* was expressed in the stamen, pistil, and ovule. Ectopic expression of *PMADS3* could lead to petals being converted into anthers [29]. In *Cyclamen persicum*, *CpAG1* and *CpAG2* were expressed in stamen and carpel. *CpAG1* is mainly involved in controlling stamen development, while *CpAG2* is mainly involved in specifying carpel identity and meristematic activity termination. Both genes work together to complete *AG* function [30,31]. In *Nicotiana benthamiana*, *NbAG* alone is able to specify stamen identity, while the C-function for pistil development needs both *NbAG* and *NbSHP*. However, the absence of the less active *NbSHP* factor only results in mild defects, and only simultaneous expression of *NbAG* and *NbSHP* can lead to floral meristem termination [32].

In other core eudicots, such as *Fagopyrum esculentum*, *FaesAG* is expressed only in stamens and pistils, and could mimic endogenous *AG* in determination of stamen and carpel identity in *ag-1* mutant *Arabidopsis* [14]. In basal eudicots, such as *Nigella damascene*, *NdAG1* determines stamen and carpel identities, while *NdAG2* participates in carpel development and floral determinacy [33]. In the basal angiosperms, such as *Magnolia wufengensis*, obvious *MawuAG1* expression was observed in the stamen, carpel, ovule, and seed, while *MawuAG2* was observed only in the stamen and carpel. Overexpression of *MawuAG1* caused sepals to be converted into carpeloid structures, with ovules and petals converted into stamen-like structures. However, overexpression of *MawuAG2* was only able to make petals convert into stamens [34]. However, *M. stellata AG*-like gene, *MastAG*, plays key role in carpel identity [35]. These data suggest a certain functional conservation and versatility of AG orthologs among different eudicot species.

In *I. verticillata*, *IlveAG* is highly expressed in stamens, pistils, and sepals of male and female flowers. Moreover, *pIlveAG* has CArG-box for the binding of MADS-box TF and AACAAA/TTTGTT motif for recognition and binding by floral homeotic APETALA2 TF [18,20]. In addition, *pIlveAG* also contains some cis-regulatory elements for recognition and binding by R2R3-MYB TFs [23,24]. In *Petunia axillaris*, the R2R3-MYB TF EOB2 is a repressor of flower bud senescence by inhibiting ethylene production and is involved in both petal and pistil maturation through regulation of primary and secondary metabolism [36]. The *DlR2R3-MYB* from *Dimocarpus longan* tends to play conserved roles in the flowering regulation and stress responses [37]. Moreover, *GAMYB* plays an important role in pollen development in both poplars and willows [38]. All these suggest that flowering and floral development may be induced by stress through the R2R3-MYB and *IlveAG* pathways. *I. verticillata* may regulate flowering or floral organ development through *IlveAG*. Additionally, *pIlveAG* can drive *GUS* to express in sepals, stamens, and carpels of transgenic *Arabidopsis*. Ectopic expression of *IlveAG* leads to early flowering in transgenic *Arabidopsis*. Moreover, *IlveAG* fully completes floral defects in the *ag-1* mutant *Arabidopsis*, and *IlveAG* can inhibit the development of sepals and petals. These results suggest that *IlveAG* can substitute the endogenous *AG* gene of *Arabidopsis ag-1* mutant to participate in the regulation of stamen and carpel development, promote the formation of floral meristem and regulate flowering, and inhibit the activity of A-class genes. *IlveAG* is a C-class gene for the regulation of flower development in *I. verticillata*. Our findings reveal the role of the *IlveAG* gene in the development of male and female flowers of *I. verticillata*, and provide more details for further analysis of flowering and fruiting of woody plants.

5. Conclusions

I. verticillata is a dioecious shrub, whereby an individual plant has either male or female flowers with vestigial organs of the opposite sex. The male flower contains pistillodes, and the female flower contains staminodes. Furthermore, the development of male and female flower buds was not synchronized. In *Arabidopsis*, *AG* plays key roles in specifying reproductive organ (stamen and pistil) identity and floral meristem determinacy, as well as repression of the A-function. However, the mechanism of male and female floral organ development in *I. verticillata* is still unknown. Here, an *AG* ortholog (*IlveAG*) and its promoter (*pIlveAG*) were isolated and characterized in *I. verticillata*. *IlveAG* is highly expressed in stamens, pistils, and sepals of male and female flowers. Moreover, *pIlveAG* could drive *GUS* to obviously express in stamens, carpels, and sepals of transgenic *Arabidopsis*. Ectopic expression of *IlveAG* leads to early flowering in transgenic *Arabidopsis*. Moreover, *IlveAG* could substitute for endogenous *AG* to rescue stamens and pistils in the *Arabidopsis ag-1* mutant. In addition, ectopic expression of *IlveAG* inhibits the development of the two outer whorls' floral organs (sepals and petals). Our findings show that *IlveAG* has a conservative C-function and plays roles in determining the identity of reproductive organs (stamen and pistil) and regulating meristem determinacy. Our results reveal the expression pattern and function of *AG* orthologs in the male and female flowers of woody plants, and provide a basis for further analysis of the flowering of *I. verticillata*.

Supplementary Materials: The following supporting information can be downloaded at: https://www.mdpi.com/article/10.3390/horticulturae10101058/s1, Figure S1: IlveAG promoter sequence; Table S1: Primers used in this study; Table S2: Information on Sequences selected for alignments and phylogenetic analyses from NCBI GenBank.

Author Contributions: J.L., writing—original draft preparation; J.L., Y.S. and X.C., methodology; Z.L., writing—review and editing, supervision. All authors have read and agreed to the published version of the manuscript.

Funding: This research received no external funding.

Data Availability Statement: All data generated or analyzed during this study are included in this published article.

Conflicts of Interest: The authors declare no conflict of interest.

References

1. Stewart, D.; Graciet, E.; Wellmer, F. Molecular and regulatory mechanisms controlling floral organ development. *FEBS J.* **2016**, *283*, 1823–1830. [CrossRef] [PubMed]
2. Theißen, G.; Melzer, R.; Rümpler, F. MADS-domain transcription factors and the floral quartet model of flower development: Linking plant development and evolution. *Development* **2016**, *143*, 3259–3271. [CrossRef] [PubMed]
3. Pelayo, M.A.; Yamaguchi, N.; Ito, T. One factor, many systems: The floral homeotic protein AGAMOUS and its epigenetic regulatory mechanisms. *Curr. Opin. Plant Biol.* **2021**, *61*, 102009. [CrossRef] [PubMed]
4. Liu, Z.; Zhang, D.; Liu, D.; Li, F.; Lu, H. Exon skipping of *AGAMOUS* homolog *PrseAG* in developing double flowers of *Prunus lannesiana* (Rosaceae). *Plant Cell Rep.* **2013**, *32*, 227–237. [CrossRef] [PubMed]
5. Tani, E.; Polidoros, A.N.; Flemetakis, E.; Stedel, C.; Kalloniati, C.; Demetriou, K.; Katinakis, P.; Tsaftaris, A.S. Characterization and expression analysis of *AGAMOUS*-like, *SEEDSTICK*-like, and *SEPALLATA*-like MADS-box genes in peach (*Prunus persica*) fruit. *Plant Physiol. Biochem.* **2009**, *47*, 690–700. [CrossRef]
6. Ma, J.; Shen, X.; Liu, Z.; Zhang, D.; Liu, W.; Liang, H.; Wang, Y.; He, Z.; Chen, F. Isolation and Characterization of *AGAMOUS*-Like Genes Associated with Double-Flower Morphogenesis in *Kerria japonica* (Rosaceae). *Plant Sci.* **2018**, *9*, 959. [CrossRef]
7. Jing, D.; Chen, W.; Xia, Y.; Shi, M.; Wang, P.; Wang, S.; Wu, D.; He, Q.; Liang, G.; Guo, Q. Homeotic transformation from stamen to petal in *Eriobotrya japonica* is associated with hormone signal transduction and reduction of the transcriptional activity of *EjAG*. *Physiol. Plant* **2020**, *168*, 893–908. [CrossRef]
8. Li, J.; Wang, L.; Chen, X.; Zeng, L.; Su, Y.; Liu, Z. Characterization of Two *AGAMOUS*-like Genes and Their Promoters from the *Cymbidium faberi* (Orchidaceae). *Plants* **2023**, *12*, 2740. [CrossRef]
9. Higo, K.; Ugawa, Y.; Iwamoto, M.; Korenaga, T. Plant cis-acting regulatory DNA elements (PLACE) database: 1999. *Nucleic Acids Res.* **1999**, *27*, 297–300. [CrossRef]

10. Clough, S.J.; Bent, A.F. Floral Dip: A Simplified Method for Agrobacterium-Mediated Transformation of *Arabidopsis thaliana*. *Plant J.* **1998**, *16*, 735–743. [CrossRef]
11. Liu, Z.; Fei, Y.; Zhang, K.; Fang, Z. Ectopic Expression of a *Fagopyrum esculentum APETALA1* Ortholog Only Rescues Sepal Development in *Arabidopsis Ap1* Mutant. *Int. J. Mol. Sci.* **2019**, *20*, 2021. [CrossRef] [PubMed]
12. You, W.; Chen, X.; Zeng, L.; Ma, Z.; Liu, Z. Characterization of *PISTILLATA*-like Genes and Their Promoters from the Distyly *Fagopyrum esculentum*. *Plants* **2022**, *11*, 1047. [CrossRef]
13. Liu, Z.; Xiong, H.; Li, L.; Fei, Y. Functional Conservation of an *AGAMOUS* Orthologous Gene Controlling Reproductive Organ Development in the Gymnosperm Species *Taxus chinensis* var. *mairei*. *J. Plant Biol.* **2018**, *61*, 50–59. [CrossRef]
14. Li, L.; Fang, Z.; Li, X.; Liu, Z. Isolation and Characterization of the C-class MADS-box Gene from the Distylous Pseudo-cereal *Fagopyrum esculentum*. *J. Plant Biol.* **2017**, *60*, 189–198. [CrossRef]
15. Neff, M.M.; Neff, J.D.; Chory, J. dCAPS, a simple technique for the genetic analysis of single nucleotide polymorphisms: Experimental applications in *Arabidopsis thaliana* genetics. *Plant J.* **1998**, *14*, 387–392. [CrossRef] [PubMed]
16. Filichkin, S.A.; Leonard, J.M.; Monteros, A.; Liu, P.P.; Nonogaki, H. A Novel Endo-β-Mannanase Gene in Tomato *LeMAN5* Is Associated with Anther and Pollen Development. *Plant Physiol.* **2004**, *134*, 1080–1087. [CrossRef]
17. Rogers, H.J.; Bate, N.; Combe, J.; Sullivan, J.; Sweetman, J.; Swan, C.; Lonsdale, D.M.; Twell, D. Functional analysis of cis-regulatory elements within the promoter of the tobacco late pollen gene g10. *Plant Mol. Biol.* **2001**, *45*, 577–585. [CrossRef]
18. de Folter, S.; Angenent, G.C. trans meets cis in MADS science. *Trends Plant Sci.* **2006**, *11*, 224–231. [CrossRef]
19. Wenkel, S.; Turck, F.; Singer, K.; Gissot, L.; Le Gourrierec, J.; Samach, A.; Coupland, G. CONSTANS and the CCAAT Box Binding Complex Share a Functionally Important Domain and Interact to Regulate Flowering of *Arabidopsis*. *Plant Cell* **2006**, *18*, 2971–2984. [CrossRef]
20. Dinh, T.T.; Girke, T.; Liu, X.; Yant, L.; Schmid, M.; Chen, X. The floral homeotic protein APETALA2 recognizes and acts through an AT-rich sequence element. *Development* **2012**, *139*, 1978–1986. [CrossRef]
21. Mena, M.; Cejudo, F.J.; Isabel-Lamoneda, I.; Carbonero, P. A Role for the DOF Transcription Factor BPBF in the Regulation of Gibberellin-Responsive Genes in Barley Aleurone. *Plant Physiol.* **2002**, *130*, 111–119. [CrossRef] [PubMed]
22. Abe, H.; Urao, T.; Ito, T.; Seki, M.; Shinozaki, K.; Yamaguchi-Shinozaki, K. Arabidopsis *AtMYC2 (bHLH)* and *AtMYB2 (MYB)* function as transcriptional activators in abscisic acid signaling. *Plant Cell* **2003**, *15*, 63–78. [CrossRef]
23. Urao, T.; Yamaguchi-Shinozaki, K.; Urao, S.; Shinozaki, K. An *Arabidopsis* myb homolog is induced by dehydration stress and its gene product binds to the conserved MYB recognition sequence. *Plant Cell* **1993**, *5*, 1529–1539. [PubMed]
24. Agarwal, M.; Hao, Y.; Kapoor, A.; Dong, C.H.; Fujii, H.; Zheng, X.; Zhu, J.K. A R2R3 type MYB transcription factor is involved in the cold regulation of CBF genes and in acquired freezing tolerance. *J. Biol. Chem.* **2006**, *281*, 37636–37645. [CrossRef]
25. Dreni, L.; Kater, M.M. MADS reloaded: Evolution of the AGAMOUS subfamily genes. *New Phytol.* **2014**, *201*, 717–732. [CrossRef] [PubMed]
26. Dreni, L.; Pilatone, A.; Yun, D.; Erreni, S.; Pajoro, A.; Caporali, E.; Zhang, D.; Kater, M.M. Functional analysis of all AGAMOUS subfamily members in rice reveals their roles in reproductive organ identity determination and meristem determinacy. *Plant Cell* **2011**, *23*, 2850–2863. [CrossRef]
27. Lv, S.; Du, X.; Lu, W.; Chong, K.; Meng, Z. Two *AGAMOUS*-like MADS-box genes from *Taihangia rupestris* (Rosaceae) reveal independent trajectories in the evolution of class C and class D floral homeotic functions. *Evol. Dev.* **2007**, *9*, 92–104.
28. Kitahara, K.; Hibino, Y.; Aida, R.; Matsumoto, S. Ectopic expression of the rose *AGAMOUS*-like MADS-box genes *'MASAKO C1* and *D1'* cause similar homeotic transformation of sepal and petal in Arabidopsis and sepal in *Torenia*. *Plant Sci.* **2004**, *166*, 1245–1252. [CrossRef]
29. Heijmans, K.; Ament, K.; Rijpkema, A.S.; Zethof, J.; Wolters-Arts, M.; Gerats, T.; Vandenbussche, M. Redefining C and D in the petunia ABC. *Plant Cell* **2012**, *24*, 2305–2317. [CrossRef]
30. Tanaka, Y.; Oshima, Y.; Yamamura, T.; Sugiyama, M.; Mitsuda, N.; Ohtsubo, N.; Ohme-Takagi, M.; Terakawa, T. Multi-petal cyclamen flowers produced by AGAMOUS chimeric repressor expression. *Sci. Rep.* **2013**, *3*, 2641. [CrossRef]
31. Tanaka, Y.; Yamamura, T.; Terakawa, T. Identification and expression analysis of the *Cyclamen persicum* MADS-box gene family. *Plant Biotechnol.* **2011**, *28*, 167–172. [CrossRef]
32. Fourquin, C.; Ferrándiz, C. Functional analyses of AGAMOUS family members in *Nicotiana benthamiana* clarify the evolution of early and late roles of C-function genes in eudicots. *Plant J. Cell Mol. Biol.* **2012**, *71*, 990–1001. [CrossRef] [PubMed]
33. Wang, P.; Liao, H.; Zhang, W.; Yu, X.; Zhang, R.; Shan, H.; Duan, X.; Yao, X.; Kong, H. Flexibility in the structure of spiral flowers and its underlying mechanisms. *Nat. Plants* **2015**, *2*, 15188. [CrossRef] [PubMed]
34. Ma, J.; Deng, S.; Jia, Z.; Sang, Z.; Zhu, Z.; Zhou, C.; Ma, L.; Chen, F. Conservation and divergence of ancestral AGAMOUS/SEEDSTICK subfamily genes from the basal angiosperm *Magnolia wufengensis*. *Tree Physiol.* **2020**, *40*, 90–107. [CrossRef] [PubMed]
35. Zhang, B.; Liu, Z.; Ma, J.; Song, Y.; Chen, F. Alternative splicing of the *AGAMOUS* orthologous gene in double flower of *Magnolia stellata* (Magnoliaceae). *Plant Sci.* **2015**, *241*, 277–285. [CrossRef]
36. Chopy, M.; Binaghi, M.; Cannarozzi, G.; Halitschke, R.; Boachon, B.; Heutink, R.; Bomzan, D.P.; Jäggi, L.; van Geest, G.; Verdonk, J.C.; et al. A single MYB transcription factor with multiple functions during flower development. *New Phytol.* **2023**, *239*, 2007–2025. [CrossRef]

37. Chen, Q.; Zhang, X.; Fang, Y.; Wang, B.; Xu, S.; Zhao, K.; Zhang, J.; Fang, J. Genome-wide identification and expression analysis o the R2R3-MYB transcription factor family revealed their potential roles in the flowering process in longan (*Dimocarpus longan* *Front. Plant Sci.* **2022**, *13*, 820439. [CrossRef]
38. Zhou, F.; Chen, Y.; Wu, H.; Yin, T. Genome-Wide Comparative Analysis of R2R3 MYB Gene Family in *Populus* and *Salix* an Identification of Male Flower Bud Development-Related Genes. *Front. Plant Sci.* **2021**, *12*, 721558. [CrossRef]

horticulturae

Article

Phenotypic and Genetic Diversity Analysis of 18 Ornamental Strawberries

Chaocui Nong [1,2], Jiayi Hou [1,2], Jin He [1,2], Yanju Zheng [1,2], Shugen Yang [1,2], Lai Jiang [1,2], Qian Xie [1,2], Wei Wang [1,2], Jinghua Wu [1,2], Qingxi Chen [1,2,*] and Lixiang Miao [1,2,*]

1 College of Horticulture, Fujian Agriculture and Forestry University, Fuzhou 350002, China; chaocuinong2024@163.com (C.N.); 18974212767@163.com (J.H.); hejin0176@163.com (J.H.); 13164419546@163.com (Y.Z.); 14769614040@163.com (S.Y.); jianglai3605@163.com (L.J.); xieq0416@163.com (Q.X.); mykuangwen@hotmail.com (W.W.); wjhham@163.com (J.W.)

2 Key Laboratory of Ministry of Education for Genetics, Breeding and Multiple Utilization of Crops, Fujian Agriculture and Forestry University, Fuzhou 350002, China

* Correspondence: cqx0246@fafu.edu.cn (Q.C.); mlx.123@163.com (L.M.)

Abstract: The red-flowered strawberry is a novel ornamental plant. This study aimed to assess the genetic diversity among ornamental strawberry germplasm resources. In this investigation, 17 red-flowered strawberry germplasms and 1 white-flowered strawberry germplasm were analyzed for genetic diversity and genetic relationships using a combination of phenotypic data, inter-simple sequence repeats (ISSR), and conserved DNA-derived polymorphism (CDDP) molecular markers. The results indicated that the 18 strawberry germplasms exhibited significant variability and genetic diversity at both phenotypic and molecular levels. The clustering results revealed notable differences between phenotypic clustering and molecular marker clustering, while the ISSR and CDDP markers grouped into broadly similar clusters. We further consolidated the ISSR and CDDP marker data to conduct the cluster analysis and population structure analysis of the 18 strawberry germplasms. The cluster analysis classified these germplasms into four clusters at a genetic similarity coefficient of 0.77. The population structure analysis categorized the germplasms into three groups, with 88.89% exhibiting a Q value $\geq$ 0.6, and 11.11% demonstrating a Q value < 0.6. This finding suggests that the genetic background of the 18 strawberry germplasms is relatively homogeneous. Notably, 'Summer Breeze-Rose' and 'Summer Breeze-Cherry' possess relatively complex genetic backgrounds (Q < 0.6). Furthermore, the floral, foliar, and plant traits of both germplasms display significant ornamental value and can serve as vital resources for the development and utilization of ornamental strawberries, as well as for the selection and breeding of new varieties.

Keywords: red-flowered strawberry; phenotypic trait; ISSR; CDDP; genetic diversity; population structure

1. Introduction

Strawberry (*Fragaria ananassa* Duch.), often referred to as the 'queen of fruits', is a perennial herb belonging to the genus *Fragaria* within the Rosaceae family. Its widespread cultivation globally can be attributed to its rich nutritional value and significant economic benefits [1,2]. The red-flowered strawberry represents a novel addition to the strawberry family, resulting from the hybridization of *F. ananassa* (2n = 8x = 56) and *Potentilla palustris* (2n = 6x = 42), and serves as a new type of ornamental flower [3]. Since the birth of the world's first red-flowered strawberry variety, 'Pink Panda', in 1989 [4], researchers from various countries have focused on enhancing ornamental traits—such as flower size, petal color, number of petals, and petal shape—alongside enhancing resistance and fruit quality, including large size and high quality in varieties such as 'Roman', 'Summer Breeze-Rose', 'Summer Breeze-Cherry', and so on [5]. In addition to red-flowered strawberries, certain white-flowered varieties, such as 'Tokun', exhibit remarkable ornamental qualities [6].

This cultivar is distinguished by its prolific flowering, large flower diameter, deep double petals, and flowers above the leaf surface, making it a valuable parental source for the development of new ornamental strawberry varieties. Now, numerous new varieties of red-flowered strawberries have been bred globally [5]. However, there are notable differences in ornamental properties among germplasms. Therefore, understanding the specific traits, genetic diversity, and genetic relationships of these germplasm resources is crucial for effective breeding.

Genetic diversity is the foundation of biological diversity and serves as a driving force behind the stability and ongoing evolution of species. Various methods for studying genetic diversity include morphological, cytological, and molecular marker techniques [7]. Among these methods, morphological markers are the simplest and most convenient, capable of revealing the extent of genetic variation and, to some degree, uncovering valuable germplasms. However, morphological traits are easily influenced by ecological conditions [8]. In contrast, molecular markers are unaffected by ecological factors and can directly identify inter-individual differences at the DNA level. This approach offers several advantages, including high specificity, high accuracy, small sample volume requirements, and rapid analyses [9]. In recent years, molecular markers have been extensively applied in studies of plant genetic diversity, assessments of population structure, analyses of genetic relationships, and the construction of core collections [10]. Molecular markers serve as an effective complement to morphological markers. The combination of molecular and morphological markers enables a more direct and accurate identification of the genetic basis underlying plant genetic variation, thereby providing a technical foundation for the evaluation of germplasm resources and multi-target selection [11]. Research methodologies that combine phenotypic and molecular markers have been successfully applied to various species, including *Rosa damascena* [12], *Camellia japonica* [13], *Clematis macropetala* [14,15], and *Chrysanthemum morifolium* [16].

Inter-simple sequence repeats (ISSRs) can amplify inter-microsatellite sequences at multiple loci across the genome, and are capable of detecting both microsatellite and inter-microsatellite polymorphisms without prior knowledge of the DNA sequences. This method is rapid, stable, highly reproducible, and cost-effective [17]. ISSR has been widely utilized to analyze the genetic diversity and relationships among various plant species, including *F. ananassa* [18,19], *Cerasus* spp. [20], and *Ficus carica* L. [21]. A conserved DNA-derived polymorphism (CDDP) is a molecular marker technique for target genes, wherein primers are specifically designed for a gene or a characteristic of a gene family in plants [22]. The resulting amplification products are typically more closely associated with the target gene. The CDDP technique is characterized by its ease of operation, low cost, and abundance of polymorphisms, and it has been extensively employed to investigate the genetic diversity of various plants, including *Paeonia suffuticosa* [23], *Camellia japonica* [13], *Rosa rugosa* [24], *Pistacia vera* L. [25], and *Musa* L. [26]. However, its application in red-flowered strawberry research has not yet been reported. ISSR molecular marker technology aims to amplify the intronic regions of the genome, which is stochastic. In contrast, CDDP molecular marker technology serves as a purposive marker, capable of amplifying bands associated with functional genes, thereby revealing the polymorphisms within these genes. The combination of ISSR and CDDP markers, which amplify distinct regions of the genome, facilitates the establishment of more reliable genetic relationships among germplasms. This approach has been effectively applied to the analysis of genetic diversity, both among and within populations of plants, such as *Iris aucheri* [27] and *Astragalus* L [28]. These studies have significant implications for the development and utilization of plant resources, the conservation of valuable germplasms, and the advancement of breeding programs.

In this study, we integrated phenotypic traits with ISSR and CDDP molecular markers to analyze the genetic diversity and relationships among 18 strawberry germplasms. This approach not only facilitates the identification of superior ornamental traits within the strawberry germplasm, but also offers significant references into its evolutionary development. Furthermore, it provides a foundational basis for the innovation of ornamental strawberry germplasm and the selection of specific resources for practical applications.

2. Materials and Methods

2.1. Test Materials

The test materials comprised 18 strawberry germplasms, including 17 red-flowered varieties (lines) and 1 white-flowered variety (Table 1). Among these, eight were sourced from China, one from the United Kingdom, eight from the Netherlands, and the white-flowered variety was sourced from Japan. On 8 October 2023, the stolon seedlings were planted in gallon pots with an upper caliber of 15 cm, a lower caliber of 11 cm, and a height of 15 cm. A total of 30 plants were planted for each germplasm, all of which were placed in the horticultural facility greenhouse of Fujian Agriculture and Forestry University in China for cultivation. Fertilizer, water, and pest control were managed according to conventional methods.

Table 1. Test materials and their sources.

No.	Name	Code Name	Source	No.	Name	Code Name	Source
1	Fenyun	FYZ	China	10	Summer Breeze-Rose	MGH	Holland
2	140	SOZ	China	11	Summer Breeze-Cherry	YHH	Holland
3	243	SCZ	China	12	Gasana	JSH	Holland
4	245	SWZ	China	13	Tarpan	YMH	Holland
5	246	SLZ	China	14	Roman	LMH	Holland
6	Zijinhong	JHZ	China	15	Frisan	FLH	Holland
7	Zijindaiyu	DYZ	China	16	Bubby Ann	BSH	Holland
8	Toscana	TCS	China	17	Merlan	MLH	Holland
9	Pink Panda	PPY	England	18	Tokun	TXR	Japan

2.2. Test Methods

2.2.1. Measurement of Phenotypic Traits

In accordance with the "Descriptors and Data Standard for Strawberry (*Fragaria* spp.)" [29], a total of 15 plants were randomly selected from each germplasm between late November 2023 and late February 2024. Observations of flower and plant correlation traits were conducted during the flower blooming date, while leaf correlation traits were observed during the late flower blooming date. Seventeen quantitative traits were assessed, including leaf length, leaf width, leaf thickness, petiole length, petiole diameter, plant height, size of plant, inflorescence stalk length, inflorescence stalk thickness, pedicel length, pedicel diameter, flower petal number, stamen number, calyx size, flower petal length, and flower petal width. Additionally, sixteen quality traits were observed through naked-eye observation, and values were assigned to these quality traits for subsequent analysis (Table 2).

Table 2. Quality trait assignment table.

Trait	Assignment of Each Character
Foliar state	Spoon = 1, Edging up = 2, Level = 3, Lever and top down = 4, Edging down = 5
Leaf color	Yellow green = 1, Green = 2, Dark green = 3, Blue green = 4
Leaf shape	Round = 1, Ellipse = 2, Rhombus = 3, Oval = 4, Upside-down oval = 5
Shape of the incisions of the leaf margin	Serrate = 1, Crenate = 2
Leaf texture	Supple = 1, Coriaceous and coarse = 2, Coriaceous and smooth = 3
Hair on the right side of the leaf	Vertical = 1, Plagiotropic = 2, Level-stick = 3
Hair on the back side of the leaf	Vertical = 1, Plagiotropic = 2, Level-stick = 3
Petiole color	Yellow-green = 1, Purple-red = 2
Petiole hair	Vertical = 1, Plagiotropic = 2, Level-stick = 3

Table 2. *Cont.*

Trait	Assignment of Each Character
Position of inflorescence relative to foliage	Beneath = 1, Level width = 2, Above = 3
Inflorescence architecture	Vertical = 1, Plagiotropic = 2
Flower color	White = 1, Light pink = 2, Pink = 3, Red = 4, Dark red = 5
Petal shape	Oblate = 1, Round = 2, Sector = 3, Oval = 4, Ellipse = 5
Position of stamen	Low = 1, Intermediate = 2, High = 3
Pedicel hair	Vertical = 1, Plagiotropic = 2, Level-stick = 3
Plant architecture	Erect = 1, Intermediate = 2, Prostrate = 3

2.2.2. DNA Extraction and Its Quality Testing

The unfolded young leaves of 18 strawberry germplasms were weighed, and total DNA was extracted from these varieties using the Biospin Omni Plant Genomic DNA Extraction Kit (Hangzhou Bioer Technology Co., Ltd., Hangzhou, China). The concentration and optical density (OD) values of the extracted DNA were measured using a UV spectrophotometer at wavelengths of 260 nm and 280 nm, respectively. The purity of the DNA was further verified through agarose gel electrophoresis. Subsequently, DNA samples of acceptable quality were uniformly diluted to a concentration of 20 ng/µL for subsequent PCR amplification assays.

2.2.3. ISSR and CDDP Primer Screening

The study utilized 100 ISSR universal primers published by the University of British Columbia (UBC) and 35 CDDP primers [22,30]. These primers were synthesized by Fuzhou Shangya Biotechnology Co. in Fuzhou, China. DNA from two strawberry varieties, '245' and 'Tokun', which display notable phenotypic differences, served as templates. Primers that exhibited strong amplification, a clear background, and high polymorphism number were selected for the PCR amplification of 18 strawberry varieties (lines) (Table 3).

Table 3. Twenty-five ISSR and seventeen CDDP primers in this study.

Primer Type	Primer Name	Primer Sequences 5′-3′	Tm (°C)
ISSR	UBC 807	AGAGAGAGAGAGAGAGT	53.0
	UBC 808	AGAGAGAGAGAGAGAGC	55.0
	UBC 810	GAGAGAGAGAGAGAGAT	53.0
	UBC 811	GAGAGAGAGAGAGAGAC	55.0
	UBC 812	GAGAGAGAGAGAGAGAA	53.0
	UBC 816	CACACACACACACACAT	53.0
	UBC 823	TCTCTCTCTCTCTCTCC	55.0
	UBC 824	TCTCTCTCTCTCTCTCG	53.0
	UBC 827	ACACACACACACACACG	53.0
	UBC 834	AGAGAGAGAGAGAGAGYT	55.8
	UBC 835	AGAGAGAGAGAGAGAGYC	55.0
	UBC 836	AGAGAGAGAGAGAGAGYA	52.3
	UBC 840	GAGAGAGAGAGAGAGAYT	53.0
	UBC 844	CTCTCTCTCTCTCTCTRC	52.3
	UBC 845	CTCTCTCTCTCTCTCTRG	55.0
	UBC 847	CACACACACACACACARC	55.0
	UBC 848	CACACACACACACACARG	55.0
	UBC 851	GTGTGTGTGTGTGTGTYG	54.3
	UBC 853	TCTCTCTCTCTCTCTCRT	52.0
	UBC 855	ACACACACACACACACYT	57.0
	UBC 857	ACACACACACACACACYG	55.0
	UBC 864	ATGATGATGATGATGATG	51.0
	UBC 868	GAAGAAGAAGAAGAAGAA	50.0
	UBC 874	CCCTCCCTCCCTCCCT	60.8
	UBC 880	GGAGAGGAGAGGAGA	55.0

Table 3. *Cont.*

Primer Type	Primer Name	Primer Sequences 5′-3′	Tm (°C)
CDDP	CHI1	GCCGTTGGAGCTACAC	50.0
	DFR1	GATCCTGCCTGAGCAAGG	54.3
	ERF3	TGGCTSGGCACSTTCGA	53.0
	F3H1	GGGAGAAGCTGTGCG	56.0
	F3H2	GGTGGCCGGATAAGCCGG	52.3
	F3H3	CGGCTGCAACCCAGTAACG	54.3
	WRKY-R1	GTGGTTGTGCTTGCC	50.0
	WRKY-R2	GCCCTCGTASGTSGT	50.0
	WEKYR-2B	TGSTGSATGCTCCCG	55.0
	ANS1	GGCCTGCAGCTCTTCT	56.0
	ANS2	GCGTCCCCAACTCGATC	59.0
	MYB1	GGCAAGGGCTGCCGC	59.0
	MYB2	GGCAAGGGCTGCCGG	59.0
	KNOX-1	AAGGGSAAGCTSCCSAAG	53.0
	KNOX-2	CACTGGTTGGGAGCTSCAC	59.0
	MADS-1	ATGGGCCGSGGCAAGGTGC	64.0
	MADS-4	CTSTGCGACCGSGAGGTG	59.0

2.2.4. PCR Amplification and Product Detection

PCR amplification was conducted in a total reaction volume of 25 µL, comprising 2.5 µL of DNA template, 1.25 µL of primers, 12.5 µL of 2× SanTaq Fast PCR Mix premix (Vazyme Biotech Co., Ltd., Nanjing, China), and 8.75 µL of double-distilled water (ddH_2O). The PCR amplification program consisted of the following steps: initial denaturation at 95 °C for 5 min; denaturation at 95 °C for 15 s; annealing at the annealing temperature (Tm) for 15 s; elongation at 72 °C for 30 s; followed by 35 cycles; a final extension at 72 °C for 6 min; and concluding the reaction at 12 °C for storage. To evaluate the amplification product, 10 µL was detected using electrophoresis on a 1.5% gel (110 V for 30–40 min), which was subsequently imaged with a gel imager and photographed for documentation.

2.3. *Statistical Analysis*

The variance, mean, standard deviation, and coefficient of variation were calculated for each quantitative trait using IBM SPSS Statistics 26. The leaves were scanned using an HP Scanjet G4050 scanner (China Hewlett-Packard Co., Ltd, Beijing, China) and flowers were photographed with a Canon EOS M5 camera (Canon China Co., Ltd., Beijing, China). The isometric classification method was applied to categorize the quantitative traits into ten levels based on the mean (X) and standard deviation (S). Specifically, the first level was defined as $Xi < X - 2S$, with each subsequent level increasing by 0.5S, and the tenth level was defined as $Xi \geq X + 2S$. The relative frequency of each level (*Pi*) was then calculated, followed by the determination of the genetic diversity index using the Shannon–Wiener diversity index (H') according to the following formula:

$$H' = -\sum_{i=1}^{n} Pi\text{In}Pi, \quad (1)$$

'*Pi*' represents the proportion of materials within the level of a trait relative to the total number of materials, while 'In' denotes the natural logarithm. Additionally, the distribution frequencies of the 18 quality traits and the Shannon–Wiener information index were calculated [13,31]. Correlation heatmaps for quantitative traits, percentage stacked histograms for quality traits, and clustering maps for phenotypic traits were generated using Origin 2024.

The bands were counted based on the PCR amplification results of ISSR and CDDP markers. Clear bands at the same site were recorded as '1', while the absence of bands was recorded as '0', resulting in the establishment of a '0–1' data matrix. The number of amplified bands, the number of polymorphic bands, and the percentage of polymorphic loci for each primer were calculated using Excel. Genetic similarity coefficients, the number of observed alleles, the number of effective alleles, Nei's diversity index (*H*), and Shannon's diversity index (*I*) for the 18 strawberry varieties (lines) were computed using PopGene 1.32 software [26]. Unweighted Pair Group Method Using Arithmetic Average (UPGMA) cluster analysis was performed based on DICE similarity coefficients using NTSYSpc 2.10e software [26,32,33]. Using the ISSR + CDDP data, the K value was determined with Structure 2.3.4 software. Subsequently, the corresponding Q value for each strawberry germplasm was calculated to analyze the population genetic structure of the strawberry germplasm resources and ascertain the genetic composition of each germplasm resource. The layout of the resulting image was performed using Adobe Photoshop 2023.

3. Results

3.1. Phenotypic and Genetic Diversity Analysis of 18 Strawberry Germplasms

3.1.1. Diversity Analysis of Quantitative Traits

The observations and statistical analyses of the quantitative traits of 18 strawberry germplasms revealed that all 17 quantitative traits exhibited rich diversity across the 18 strawberry germplasms (Table 4, Figure 1). The coefficients of variation ranged from 16% to 52%, with the highest coefficients observed for inflorescence stalk length and the lowest for flower petal width. Notably, 14 of the traits demonstrated coefficients of variation exceeding 20%, indicating substantial differences among the cultivars. The degree of variation for flower diameter, flower petal length, and flower petal width was moderate, with coefficients of variation of 19%, 17%, and 16%, respectively, suggesting relatively stable performance among these varieties. The Shannon–Wiener diversity index (H') for the 17 quantitative traits ranged from 1.498 to 2.080, highlighting significant genetic differences and rich diversity among the strawberry germplasm resources under investigation. The Shannon–Wiener diversity index (H') for flower diameter was the highest (2.080), followed closely by pedicel diameter and plant height, which had diversity indices of 2.070 and 2.068, respectively. Conversely, the flower petal number exhibited the lowest diversity index (1.498), followed by stamen numbers and leaf length, which had diversity indices of 1.911 and 1.918, respectively. It is noteworthy that 'Summer Breeze-Rose' and 'Summer Breeze-Cherry' exhibited deeply double flowers, characterized by petal numbers ranging from 7 to 15, and 10 to 17 petals, respectively. The mean petal numbers for 'Summer Breeze-Rose' and 'Summer Breeze-Cherry' were 11.8 and 13.73, respectively, both of which were significantly greater than those observed in other germplasms.

Table 4. Diversity analysis of the 17 quantitative traits.

Quantitative Character	Range	Mean ± Standard Deviation	Variable Coefficient (%)	H'
Leaf length/cm	2.30~8.50	4.70 ± 1.22	25.96	1.918
Leaf width/cm	2.40~8.20	4.45 ± 1.12	25.17	1.972
Leaf thickness/mm	0.12~0.52	0.26 ± 0.06	23.08	2.013
Petiole length/cm	2.60~15.10	7.56 ± 2.46	32.54	2.049
Petiole diameter/mm	0.96~3.05	1.63 ± 0.37	22.30	2.009
Plant height/cm	5.30~19.60	11.26 ± 2.89	25.67	2.068
Size of plant/cm	6.98~27.90	14.78 ± 3.81	25.78	2.027
Inflorescence stalk length/cm	0.07~12.20	4.38 ± 2.29	52.28	2.016
Inflorescence stalk thickness/mm	0.58~2.63	1.75 ± 0.41	23.43	2.035
Pedicel length/cm	0.99~8.50	3.48 ± 1.60	45.98	1.974
Pedicel diameter/mm	0.46~1.69	1.12 ± 0.26	23.21	2.070
Flower diameter/cm	1.45~3.87	2.74 ± 0.51	18.61	2.080
Flower petal number	4.00~17.00	6.94 ± 2.73	39.33	1.498
Stamen numbers	12.00~62.00	26.16 ± 6.18	23.62	1.911
Calyx size/cm	0.93~3.93	2.16 ± 0.58	26.85	2.021
Flower petal length/cm	0.82~1.87	1.21 ± 0.20	16.53	1.964
Flower petal width/cm	0.89~2.00	1.31 ± 0.21	16.03	2.047

Figure 1. Flowers and leaves of the 18 strawberry germplasm resources. Please refer to Table 1 for the code names of the 18 strawberry germplasms.

Significant correlations were observed among 11 of the 17 quantitative traits, including highly significant positive correlations between leaf length and leaf width, petiole length, and size of plant, with correlation coefficients of 0.8720, 0.6347, and 0.7317, respectively (Figure 2). Furthermore, there were highly significant positive correlations between leaf width and petiole diameter, leaf thickness and pedicel diameter, petiole length and plant height, petiole diameter and pedicel diameter, stamen number and calyx size, as well as flower petal length and flower petal width, with correlation coefficients of 0.7544, 0.6698, 0.8493, 0.7523, 0.6161, and 0.6409.

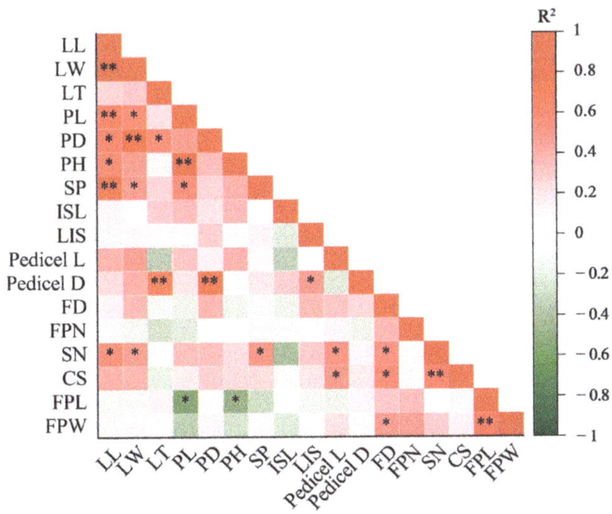

Figure 2. Correlation analysis was conducted among 17 quantitative traits of 18 strawberry germplasms. The asterisk (*) denotes significance at $p \leq 0.05$, while the double asterisk (**) indicates significance at $p \leq 0.01$. The analyzed traits include the following: LL (leaf length), LW (leaf width), LT (leaf thickness), PL (petiole length), PD (petiole diameter), PH (plant height), SP (size of plant), ISL (inflorescence stalk length), LIS (inflorescence stalk thickness), Pedicel L (pedicel length), Pedicel D (pedicel diameter), FD (flower diameter), FPN (flower petal number), SN (stamen number), CS (calyx size), FPL (flower petal length), and FPW (flower petal width).

3.1.2. Diversity Analysis of Quality Traits

We identified a total of 16 quality traits in 18 strawberry germplasms, each exhibiting varying degrees of variability. The Shannon–Wiener diversity index (H') for these 16 quality traits ranged from 0.215 to 1.480, with the highest diversity index observed for flower color (1.480) and the lowest for petiole hair (0.215). Notably, the Shannon–Wiener diversity index (H') for flower color, leaf shape, pedicel hair, and leaf texture exceeded 1, indicating a high level of genetic diversity among the 18 strawberry varieties (lines) (Figure 3). Notably, '140', '243', '245', 'Frisan', 'Roman', 'Gasana', 'Summer Breeze-Rose', and 'Summer Breeze-Cherry' exhibit compact plants and attractive flower shapes. The flowers are vividly colored, with over two-thirds positioned above the foliage, enhancing their visibility.

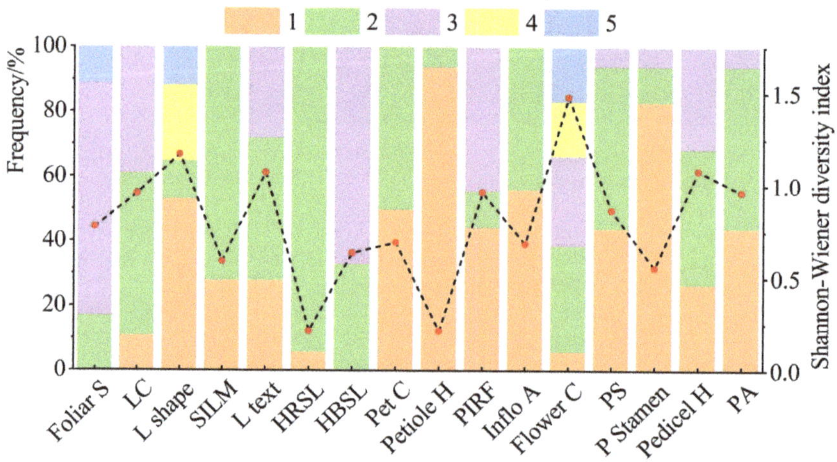

Figure 3. Frequency distribution of variant types and the Shannon–Wiener diversity index of 16 quality traits in 18 strawberry germplasms. These red dots represent the Shannon–Wiener diversity index for each of the 16 quality traits. Numbers 1 to 5 represent the values assigned to various quality traits. Please refer to Table 1 for the code names of the 18 strawberry germplasms. The analyzed traits include the following: Foliar S (foliar state), LC (leaf color), L shape (leaf shape), SILM (shape of incisions of leaf margin), L text (leaf texture), HRSL (hair on the right side of the leaf), HBSL (hair on the back side of the leaf), Pet C (petiole color), Petiole H (petiole hair), PIRF (position of inflorescence relative to foliage), Inflo A (inflorescence architecture), Flower C (flower color), PS (petal shape), P Stamen (position of stamen), Pedicel H (pedicel hair), and PA (plant architecture).

3.2. *Genetic Diversity Analysis of 18 Strawberry Germplasms Based on ISSR and CDDP Markers*

3.2.1. DNA Extraction and Primer Screening Results

Genomic DNA from 18 strawberry varieties (lines) was extracted using a plant genomic DNA extraction kit. Electrophoresis revealed that the bands were clear and free from dispersion. The DNA concentrations of the 18 strawberry varieties ranged from 78 to 114 ng/μL, with OD260/OD280 ratios between 1.7 and 2.0. These quality tests confirmed that the DNA met the necessary standards for use in subsequent experiments. The genomic DNA of red-flowered ('245') and white-flowered ('Tokun') strawberry germplasms was utilized to screen 25 ISSR and 17 CDDP primers, which exhibited good repetitiveness and polymorphisms for the current experiment. Part of the amplification electropherogram results of the primer are shown in Figure 4.

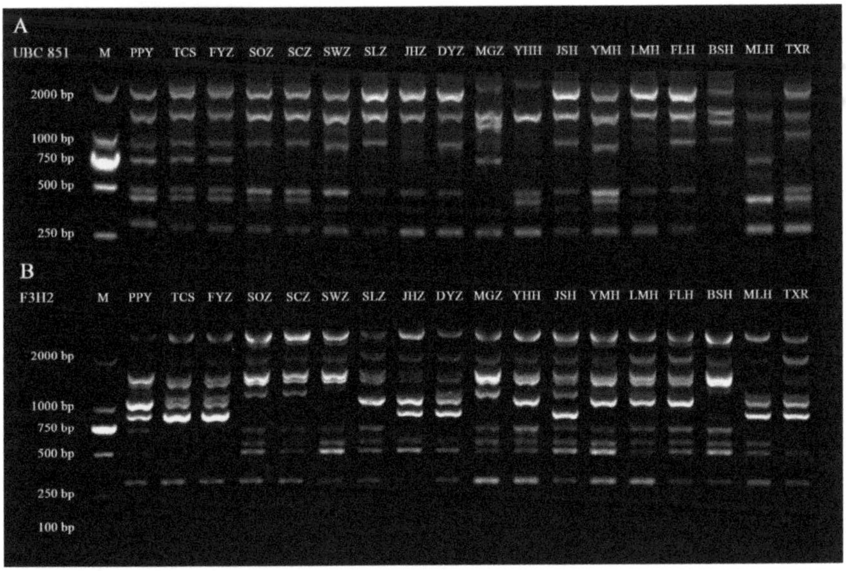

Figure 4. PCR amplification electropherogram of primer UBC 851 (**A**) and primer F3H2 (**B**). Please refer to Table 1 for the code names of the 18 strawberry germplasms.

3.2.2. Amplification Results and Polymorphism Analysis of ISSR and CDDP Molecular Markers

The PCR amplification of 18 strawberry germplasms using 25 ISSR primers resulted in a total of 188 clear bands, of which 145 were polymorphic, representing 75.91% of the total loci. The percentage of polymorphic loci exceeded 66.67% for all primers, with the exception of primers UBC 823 (40.00%) and UBC 864 (42.86%), which displayed a lower percentage of polymorphic loci (Table 5). Notably, the percentage of polymorphic loci for primers UBC 807, UBC 808, UBC 811, UBC 824, UBC 827, UBC 834, UBC 835, UBC 853, UBC 855, and UBC 880 exceeded 80%. Similarly, the PCR amplification of the same 18 strawberry germplasms using 17 CDDP primers resulted in a total of 159 clear bands, with 130 being polymorphic, accounting for 81.02% of the total loci. The percentage of polymorphic loci exceeded 80% for all primers except for CHI1 (57.14%), F3H2 (66.67%), ANS1 (78.57%), KNOX-2 (75.00%), and MADS-4 (66.67%) (Table 5). Additionally, a total of 347 bands were amplified by both ISSR and CDDP primers, with 275 of these being polymorphic, yielding an average percentage of polymorphic loci of 77.98%. These results indicate that CDDP markers amplified a greater number of polymorphic bands with fewer primers, suggesting that CDDP markers exhibit stronger polymorphism in the 18 strawberry germplasms.

The parameters of the number of alleles (*Na*), the number of effective alleles (*Ne*), Nei's diversity index (*H*), and Shannon's diversity index (*I*) for strawberry germplasms were calculated using PopGen32. The results from the ISSR molecular marker analysis indicated that the average *Na* among the 18 strawberry germplasm samples was 1.7539, the average *Ne* was 1.4084, the average *H* was 0.2443, and the average *I* was 0.3717 (Table 5). Additionally, the results from the CDDP molecular marker analysis indicated that the mean number of alleles (*Na*) among the 18 strawberry germplasms was 1.8175, with the mean number of effective alleles (*Ne*) at 1.4536, the mean Nei's diversity index (*H*) at 0.2701, and the mean Shannon's diversity index (*I*) at 0.4093 (Table 5. The ISSR + CDDP molecular marker analysis revealed that the average number of alleles (*Na*) among the 18 strawberry germplasm samples was 1.7796. Additionally, the average number of effective alleles (*Ne*) was 1.4267, the average Nei's diversity index (*H*) was 0.2547, and the average Shannon's diversity index (*I*) was 0.3869. These results indicate a relatively high level of genetic diversity among the 18 strawberry materials tested.

Table 5. ISSR and CDDP of primer polymorphisms.

Primer Type	Primer Name	Band Number	Polymorphic Band Number	Percentage of Polymorphic Loci (%)	Number of Alleles (*Na*)	Number of Effective Alleles (*Ne*)	Nei's Diversity Index (*H*)	Shannon's Diversity Index (*I*)
ISSR	UBC 807	6	6	100.00	2.0000	1.5168	0.3066	0.4678
	UBC 808	8	7	87.50	1.8889	1.5180	0.3038	0.4573
	UBC 810	11	8	72.73	1.7273	1.4489	0.2615	0.3891
	UBC 811	6	5	83.33	1.8333	1.3753	0.2366	0.3718
	UBC 812	8	6	75.00	1.7500	1.5302	0.3017	0.4416
	UBC 816	6	4	66.67	1.5000	1.2116	0.1296	0.2053
	UBC 823	5	2	40.00	1.6000	1.2420	0.1407	0.2232
	UBC 824	9	8	88.89	1.7778	1.4388	0.2688	0.4074
	UBC 827	7	6	85.71	2.0000	1.7139	0.3783	0.5475
	UBC 834	10	9	90.00	1.9000	1.4593	0.2741	0.4206
	UBC 835	5	4	80.00	1.6000	1.3166	0.2049	0.3142
	UBC 836	9	7	77.78	1.7778	1.4381	0.2627	0.3989
	UBC 840	6	4	66.67	1.8333	1.4536	0.2788	0.4258
	UBC 844	10	7	70.00	1.7000	1.3740	0.2333	0.3575
	UBC 845	4	3	75.00	1.5000	1.1968	0.1265	0.2014
	UBC 847	3	2	66.67	1.6667	1.5013	0.2798	0.4056
	UBC 848	9	7	77.78	1.7778	1.3838	0.2325	0.3580
	UBC 851	9	6	66.67	1.7778	1.4165	0.2551	0.3902
	UBC 853	7	6	85.71	1.8571	1.6127	0.3492	0.5098
	UBC 855	12	12	100.00	2.0000	1.5425	0.3215	0.4876
	UBC 857	8	6	75.00	1.6250	1.2861	0.1736	0.2705
	UBC 864	7	3	42.86	1.4286	1.1879	0.1217	0.1907
	UBC 868	9	6	66.67	1.7778	1.3725	0.2387	0.3727
	UBC 874	7	5	71.43	1.7143	1.3033	0.1958	0.3129
	UBC 880	7	6	85.71	1.8333	1.3683	0.2315	0.3642
	Total	188	145	-	-	-	-	-
	Mean	7.52	5.8	75.91	1.7539	1.4084	0.2443	0.3717
CDDP	CHI1	7	4	57.14	1.5714	1.4400	0.2451	0.3537
	DFR1	6	5	83.33	1.8333	1.5635	0.3251	0.4790
	ERF3	13	13	100.00	2.0000	1.4899	0.2982	0.4606
	F3H1	6	5	83.33	1.8333	1.4421	0.2685	0.4120
	F3H2	12	8	66.67	1.6667	1.3955	0.2258	0.3389
	F3H3	11	9	81.82	1.8182	1.5433	0.3126	0.4619
	WRKY-R1	7	6	85.71	1.8571	1.5819	0.3342	0.4925
	WRKY-R2	10	9	90.00	1.9000	1.5787	0.3364	0.4997
	WEKYR-2B	8	8	100.00	2.0000	1.3564	0.2423	0.3943
	ANS1	14	11	78.57	1.7857	1.3550	0.2279	0.3588
	ANS2	5	4	80.00	1.8000	1.1472	0.1185	0.2189
	MYB1	14	12	85.71	1.8571	1.4304	0.2623	0.4043
	MYB2	12	10	83.33	1.8333	1.3961	0.2449	0.3797
	KNOX-1	10	8	80.00	1.8000	1.5513	0.3117	0.4559
	KNOX-2	8	6	75.00	1.8750	1.4482	0.2600	0.3964
	MADS-1	10	8	80.00	1.8000	1.5413	0.3142	0.4629
	MADS-4	6	4	66.67	1.6667	1.4504	0.2644	0.3895
	Total	159	130	-	-	-	-	-
	Mean	9.35	7.65	81.02	1.8175	1.4536	0.2701	0.4093
ISSR + CDDP	Total	347	275	-	-	-	-	-
	Mean	8.26	6.55	77.98	1.7796	1.4267	0.2547	0.3869

3.3. Cluster Analysis of 18 Strawberry Germplasms Based on Phenotypic Traits, ISSR, CDDP, and ISSR + CDDP Markers

The cluster analysis of 18 strawberry germplasms based on 33 phenotypic traits categorized these germplasms into three major clusters (Figure 5A). Cluster I comprises five varieties (lines), namely 'Pink Panda', 'Merlan', 'Roman', 'Toscana', and 'Fenyun'. Cluster II includes nine varieties (lines), namely '140', '243', 'Tarpan', 'Bubby Ann', 'SummerBreeze-Rose', 'Summer Breeze-Cherry', '245', 'Gasana', and 'Frisan'. Cluster III contains four varieties (lines), namely '246', 'Zijinhong', 'Zijindaiyu', and 'Tokun'. Both Cluster I and Cluster III germplasms exhibit pink or white petal colors, while Cluster II displays a diverse range of petal colors, including light pink, pink, red, and dark red.

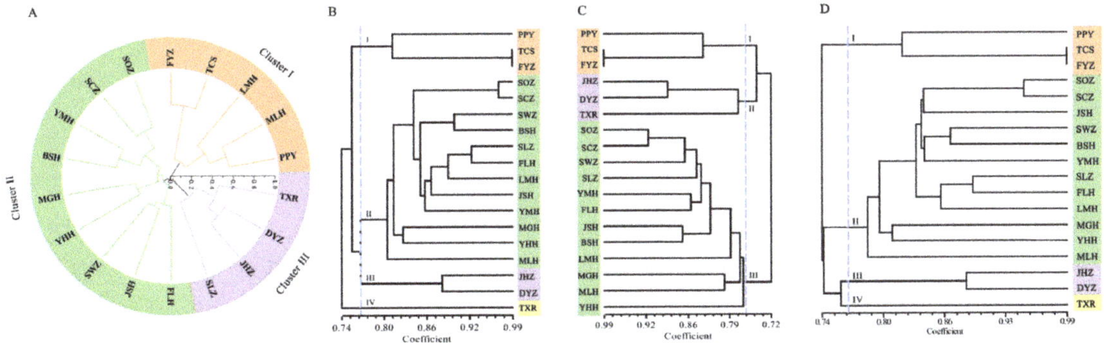

Figure 5. Cluster analysis plot of 18 strawberry germplasm resources based on phenotypic traits (**A**), and ISSR (**B**), CDDP (**C**), and ISSR + CDDP (**D**) markers. Please refer to Table 1 for the code names of the 18 strawberry germplasms.

UPGMA clustering analysis was conducted on 18 strawberry varieties, resulting in the construction of a dendrogram. Based on the ISSR molecular marker data, the 18 strawberry germplasms were categorized into four primary clusters at a similarity coefficient of 0.77 (Figure 5B). Cluster I comprises three varieties: 'Pink Panda', 'Toscana', and 'Fenyun'. Cluster II includes twelve varieties: '140', '243', '245', 'Bubby Ann', '246', 'Frisan', 'Roman', 'Gasana', 'Tarpan', 'Summer Breeze-Rose', 'Summer Breeze-Cherry', and 'Merlan'. Cluster III consists of two varieties: 'Zijinhong' and 'Zijindaiyu', while Cluster IV contains only one variety, 'Tokun'.

At a similarity coefficient of 0.77, the 18 strawberry germplasms were categorized into three major clusters based on CDDP molecular marker data (Figure 5C). Cluster I includes three species (lines) aligning with Cluster I in the ISSR molecular marker-based clustering results. Cluster II comprises 'Zijinhong', 'Zijindaiyu', and 'Tokun'. Cluster III includes 12 species (lines), which corresponds to Class II of the ISSR molecular marker-based clustering results. Notably, the phenotype-based clustering results differed significantly from those based on molecular markers. However, the clustering results obtained from ISSR and CDDP were largely consistent across the three major clusters, with discrepancies noted in the clustering outcomes of the minor clusters.

To enhance the understanding of the relationships among the test materials, this study employed a combination of ISSR and CDDP marker data to construct a clustering tree (Figure 5D). At a genetic similarity coefficient of 0.77, the 18 strawberry germplasms were classified into four major clusters based on the ISSR + CDDP molecular marker data. Cluster I comprises 'Pink Panda', 'Toscana', and 'Fenyun'; Cluster II includes '140', '243', 'Gasana', '245', 'Bubby Ann', 'Tarpan', '246', 'Frisan', 'Roman', 'Summer Breeze-Rose', 'Summer Breeze-Cherry', and 'Merlan'. Cluster III consists of 'Zijinhong' and 'Zijindaiyu', while Cluster IV is represented by 'Tokun'. The genetic similarity coefficients among the 18 strawberry germplasms, based on ISSR + CDDP markers, ranged from 0.6138 to 0.9856, with a mean value of 0.7241. Notably, 'Toscana' and 'Fenyun' exhibited the highest genetic similarity coefficient, indicating that these two varieties are the most closely related. Conversely, the lowest genetic similarity coefficients were observed between 'Tokun' and 'Bubby Ann', suggesting that these two varieties are the most distantly related.

3.4. Population Genetic Structure Analysis of 18 Strawberry Germplasms Based on ISSR + CDDP Markers

To further clarify the genetic composition of the test materials, 18 strawberry germplasm samples were analyzed for population structure using Structure software (version 2.3.4) based on ISSR + CDDP marker data. Following the method outlined by Evanno et al. [34], the change curve of ΔK was plotted against increasing K values. The analysis revealed that ΔK reached its maximum when K = 3, indicating that it is most appropriate to classify the 18 strawberry germplasms into three distinct groups (Figure 6A). The results of the population structure analysis classified the 18 strawberry germplasms into three distinct groups (Figure 6B). The first group included two germplasms, 'Toscana' and 'Fenyun', which had Q values of 0.996 and 0.991, respectively. The second group comprised nine germplasms: '140', '243', '245', '246', 'Gasana', 'Tarpan', 'Roman', 'Frisan', and 'Bubby Ann', exhibiting Q values that ranged from 0.688 to 0.989. The third group contained seven germplasms: 'Pink Panda', 'Zijinhong', 'Zijindaiyu', 'Summer Breeze-Rose', 'Summer Breeze-Cherry', 'Merlan', and 'Tokun', with Q values ranging from 0.510 to 0.991. Germplasms with a Q value of $\geq$ 0.6 are considered to have a relatively homogeneous genetic background, whereas germplasms with a Q value of <0.6 are classified as having a mixed origin [35]. Among the 18 strawberry germplasm samples analyzed, 12 exhibited a Q value greater than 0.9, representing 66.67% of the total. Additionally, 16 germplasm samples had a Q value of $\geq$0.6, accounting for 88.89%, while 'Summer Breeze-Rose' and 'Summer Breeze-Cherry' had a Q value of <0.6, corresponding to 11.11%. These findings suggest that the majority of the 18 strawberry germplasm samples possess a relatively homogeneous genetic background.

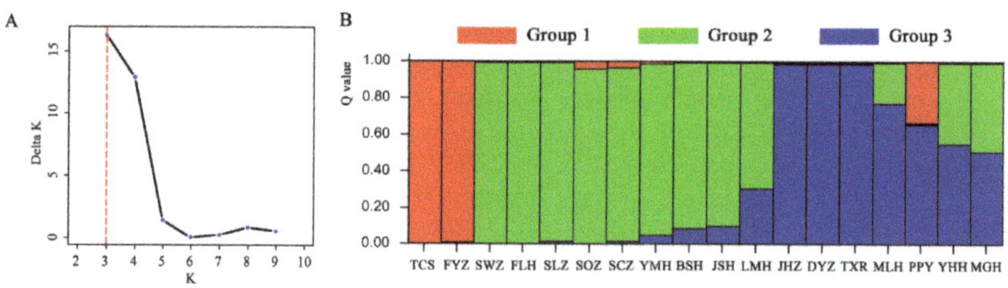

Figure 6. Trend of the rational groups' K number and estimated value ΔK based on Structure analysis (**A**). Population genetic structure analysis of 18 strawberry germplasms based on ISSR + CDDP markers (**B**). Please refer to Table 1 for the code names of the 18 strawberry germplasms.

4. Discussion

Germplasm resources serve as the foundation for breeding [36,37]. A thorough understanding of kinship distance and variation among these resources is essential for the purposeful selection of parent plants and the development of superior varieties. Consequently, analyzing the genetic diversity of germplasm resources plays a crucial role in the breeding process. Plant phenotypic diversity arises from a combination of genetic and environmental factors. Understanding the variation and diversity within red-flowered strawberry germplasm resources is a crucial step in the identification, evaluation, conservation, and utilization of these resources. When the coefficient of variation exceeds 10%, it indicates that the trait exhibits significant variability among germplasms [38–40]. Likewise, a diversity index greater than 1 signifies a high level of phenotypic trait diversity [41]. In this study, the coefficient of variation and genetic diversity of the 33 phenotypic traits investigated showed considerable differences. The coefficients of variation for the 17 quantitative traits ranged from 16% to 52%. The Shannon–Wiener diversity index for these traits ranged from 1.498 to 2.080, with a mean value of 1.981. In contrast, the coefficients of variation for the 16 quality traits ranged from 12% to 69%. The Shannon–Wiener diversity index for the quality traits ranged from 0.215 to 1.480, with a mean value of 0.761. The results indicated that the coefficients of variation for the 33 phenotypic traits exceeded 10%, with 28 of these traits exhibiting coefficients of variation greater than 20%. Furthermore, the Shannon–Wiener diversity index for 21 traits was found to be greater than 1. These findings suggest that the phenotypic traits of the test materials exhibit substantial variability and genetic diversity. Notably, the Shannon–Wiener diversity index for quantitative traits was higher than that for quality traits, a result that aligns with previous studies conducted on *Medicago ruthenica* L. [42] and *Agropyron* [43]. This observation may be attributed to the differing genetic regularity associated with quality and quantitative traits.

Plant phenotypes are influenced by environmental factors to some extent, while molecular markers can reveal genetic variation in plants at the DNA level. This approach is a stable and reliable method for genetic analysis and serves as a crucial tool in studying the genetic diversity of germplasm resources [44]. In this study, 145 and 130 polymorphic bands were amplified using 25 ISSR and 17 CDDP primers, respectively. The average percentage of polymorphic loci for these two types of markers was 75.91% and 81.02%, respectively, both of which were higher than the percentages of polymorphic loci observed in strawberry germplasms using ISSR markers, as reported by Hua et al. [18] and Zhang et al. [19]. These findings indicate that both ISSR and CDDP markers can be effectively utilized in the genetic diversity studies of red-flowered strawberry. The Nei's diversity index (H) and Shannon's diversity index (I) values for 18 samples based on ISSR markers in this study were 0.2443 and 0.3717, respectively. These values were higher than those reported in a study on *Cerasus* spp. by Liu et al. [20] using ISSR molecular markers ($H = 0.2240$; $I = 0.3623$), but lower than those found by Yang et al. [45] in their research on *Habenaria dentata* ($H = 0.2240$; $I = 0.3623$). This discrepancy may be attributed to the differences in species and the primers used. Furthermore, the average H and I values for the 18 samples based on CDDP markers were 0.2701 and 0.4093, respectively. These values were higher than those obtained from the ISSR markers, indicating that CDDP markers more effectively reflect the genetic diversity of red-flowered strawberries compared to ISSR markers. The mean Nei's diversity index (H) for the 18 strawberry germplasms, assessed using both ISSR and CDDP molecular markers, was higher than the average genetic diversity level of plants (0.22–0.23) reported by N. Hilde [46]. This finding indicates that both molecular markers effectively reflect the high level of genetic diversity present in the 18 strawberry germplasms.

In this study, 18 strawberry germplasms were clustered using phenotypic, ISSR, and CDDP markers. The results showed large differences between the phenotypic clustering and molecular marker clustering results. This difference may be due to several reasons. On the one hand, the disturbance of external factors resulted in unstable genetic expression, which led to large phenotypic differences, which is consistent with the findings of Yang et al. [47] and Zhai et al. [48]. On the other hand, it may be due to the insufficient number of morphological traits utilized in this study and the limited range of genetic differences probed by molecular markers to comprehensively reflect the genetic information of these germplasms, which resulted in the inconsistency between the results of phenotypic clustering and molecular marker clustering. In this study, 18 strawberry germplasms were classified into four and three classes based on ISSR and CDDP markers, respectively, at a genetic similarity coefficient of 0.77. The classification results of the two clustering methods were generally consistent, and only the clustering results of individual varieties differed. ISSR markers classified 'Tokun' as a separate cluster, while CDDP markers clustered 'Tokun' with 'Zijinhong' and 'Zijindaiyu'. This clustering result is justified by the fact that 'Tokun', 'Zijinhong', and 'Zijindaiyu' are all hybrids of 'Sweet Charlie' [49–51]. Two primary factors contribute to the observed differences in the results of the two clustering methods. The first factor is the variation in the binding sites of the ISSR and CDDP primers, which leads to differences in both the number and location of amplified bands. This observation aligns with the findings reported by Nawroz Tahir et al. [52]. The second factor is that the two markers employed for cluster analysis in this study amplified differing numbers of polymorphic bands; specifically, the ISSR marker displayed a greater number of polymorphic bands compared to the CDDP marker. Consequently, the ISSR marker was able to cluster 'Tokun' separately from 'Zijinhong' and 'Zijindaiyu'. To enhance the accuracy of distinguishing kinship among the germplasms, we used a combination of ISSR and CDDP markers for cluster analysis, and finally classified the 18 strawberry germplasms into four classes. However, even with the combination of the two markers, the results of the molecular markers still failed to differentiate between 'Fenyun' and 'Tuscany', suggesting that additional primers or alternative types of molecular markers may be necessary for their distinction.

Population structure analysis classified the 18 strawberry germplasms into three distinct classes, which was not entirely consistent with the UPGMA clustering results. This inconsistency arises because UPGMA clustering assumes a uniform evolutionary rate across all branches and constructs a clustering tree based on the genetic distance between pairs of germplasms. In contrast, population structure analysis utilizing the Structure software employs a Bayesian algorithm that infers population structure based on posterior probabilities by modeling the distribution of allele frequencies across various subgroups. Consequently, different algorithms may yield inconsistent results. Population structure analysis indicated that the genetic background of the 18 strawberry germplasm lines was relatively homogeneous, with 88.9% exhibiting a Q value greater than or equal to 0.6, while only 11.11% displayed a Q value less than 0.6. The two germplasms are 'Summer Breeze-Rose' and 'Summer Breeze-Cherry'. Notably, the flower petal number for these germplasms range from 7 to 15 for 'Summer Breeze-Rose' and from 10 to 17 for 'Summer Breeze-Cherry', with an average flower petal number of 11.8 and 13.73, respectively. Moreover, the floral, foliar, and plant traits of both germplasms exhibit significant ornamental value and can serve as crucial resources for the development and utilization of ornamental strawberries, as well as for the selection and breeding of new varieties.

5. Conclusions

In this study, we analyzed the genetic diversity and population structure of 18 ornamental strawberry cultivars using phenotypic, ISSR, and CDDP markers. Our findings indicate that these 18 strawberry germplasms exhibit significant variation and genetic diversity at both phenotypic and molecular levels. The clustering results revealed notable discrepancies between phenotypic clustering and molecular marker clustering, while the ISSR and CDDP markers formed similar clusters. The average genetic similarity coefficient among the 18 strawberry germplasms was notably high, at 0.7241. Population structure analysis classified the 18 strawberry germplasms into three distinct groups, with a relatively high proportion (88.89%) exhibiting a Q value ≥ 0.6. This finding suggests that the genetic background of the 18 strawberry germplasms is relatively homogeneous. The floral, foliar, and plant traits of 'Summer Breeze-Rose' and 'Summer Breeze-Cherry' demonstrate significant ornamental value and can serve as crucial resources for the development and utilization of ornamental strawberries, as well as for the selection and breeding of new varieties.

Author Contributions: Conceptualization, L.M., Q.C. and C.N.; methodology, L.M., Q.C., J.H. (Jiayi Hou) and C.N.; software, C.N., J.H. (Jiayi Hou) and J.H. (Jin He); validation, Y.Z., Q.X., W.W. and L.J.; formal analysis, C.N.; investigation, C.N. and S.Y.; resources, C.N.; data curation, C.N.; writing—original draft preparation, C.N.; writing—review and editing, C.N. and L.M.; visualization, C.N.; supervision, L.M., Q.C. and J.W.; project administration, L.M. and Q.C.; funding acquisition, L.M. and Q.C. All authors have read and agreed to the published version of the manuscript.

Funding: This research was funded by the Natural Science Foundation of Fujian Province (2023J01450), Training Funds for Core Young Scholar of Horticulture College, Fujian Agriculture and Forestry University (722022011).

Data Availability Statement: Data supporting the reported results can be requested by contacting the corresponding author. The data are not publicly available due to compliance with data protection regulations.

Acknowledgments: We gratefully acknowledge Yongchao Han (Institute of Industrial Crops, Hubei Academy of Agricultural Sciences, China) for his technical assistance.

Conflicts of Interest: The authors declare no conflicts of interest.

References

1. Wan, C.Y.; Lin, M.; Feng, L.J.; Zhi, H.H.; Chen, X.P. Progress in new cultivar breeding of strawberry in China and breeding practice. *Acta Agric. Jiangxi* **2010**, *22*, 37–39. (In Chinese)
2. Tapia, R.R.; Barbey, C.R.; Chandra, S.; Folta, K.M.; Whitaker, V.M.; Lee, S. Evolution of the *MLO* gene families in octoploid strawberry (*Fragaria* × *ananassa*) and progenitor diploid species identified potential genes for strawberry powdery mildew resistance. *Hortic. Res.* **2021**, *8*, 153. [CrossRef]
3. Asker, S. Some viewpoints on *Fragaria* × *Potentilla* intergeneric hybridization. *Hereditas* **2009**, *67*, 181–190. [CrossRef]
4. Xue, L.; Lei, J.J.; Liu, Y. Review on pink-flowered strawberry breeding. *J. Northeast Agric. Univ.* **2012**, *43*, 172–176. (In Chinese)
5. Cai, Z.; Yue, J.Y.; Wang, Y.W.; Hui, T.J.; Lei, J.J.; Xue, L. Advances in research on breeding and petal coloration mechanism of red-flowered strawberry. *J. Fruit Sci.* **2024**, *41*, 155–161. (In Chinese)
6. Xiao, G.; Guan, Z.; Zhang, S.; Wen, X.; Chen, X.; Zeng, X.; Zhang, Q.; Liu, S.; Liu, S.; Wang, Y.; et al. Genetic diversity and comprehensive evaluation of phenotypic traits in 73 germplasm resources of cultivated strawberries grown in Hubei province. *J. Fruit Sci.* **2023**, *40*, 1546–1558. (In Chinese)
7. Li, J.; Gao, G.C.; Li, B.; Li, B.; Lu, Q.H. Genetic analysis of *Prunus salicina* L. by random amplified polymorphic DNA (RAPD) and intersimple sequence repeat (ISSR). *Genet. Res.* **2022**, e63. [CrossRef]
8. Zhu, Y.Z.; Liang, D.Y.; Song, Z.J.; Tan, Y.; Guo, X.L.; Wang, D.L. Genetic diversity analysis and core germplasm collection construction of *Camellia* oleifera based on fruit phenotype and SSR data. *Genes* **2022**, *13*, 2351. [CrossRef] [PubMed]
9. Wang, G.; Cao, P.; Wei, X.M.; Hao, J.P. Applications of molecular markers in study of germplasm resources of medicinal plants. *Mod. Chin. Med.* **2019**, *21*, 1435–1444. (In Chinese)
10. Liu, Y.; Fang, X.M.; Tang, T.; Wang, Y.D.; Wu, Y.H.; Luo, J.Y.; Wu, H.T.; Wang, Y.Q.; Zhang, J.; Ruan, R.W.; et al. Inflorescence transcriptome sequencing and development of new EST-SSR markers in common Buckwheat (*Fagopyrum esculentum*). *Plants* **2022**, *11*, 742. [CrossRef] [PubMed]
11. Zhu, Y.L.; Guo, F.G. Genetic diversity analysis of *Cotoneaster coriaceus* based on phenotypic traits and ISSR marker. *J. Yunnan Agric. Univ.* **2020**, *35*, 693–699. (In Chinese)
12. Lebkiri, N.; Abbas, Y.; Iraqi, D.; Gaboun, F.; Saghir, K.; Fokar, M.; El Hamdi, I.; Bakhy, K.; Abdelwahd, R.; Diria, G. Morphological characterization and genetic diversity of mini core collection of *Rosa damascena* from Morocco. *J. Genet. Eng. Biotechnol.* **2024**, *22*, 100423. [CrossRef] [PubMed]
13. Pan, R.; Wei, Y.H.; Lai, G.D.; Que, Q.X.; Li, S.Y.; Xv, H.; Pan, H.; Chen, G.X.; Lai, C.C. Genetic diversity and association analysis of *Camellia Japonica* based on phenotypes and CDDP molecular markers technique. *Mol. Plant Breed.* **2023**, 1–25. Available online: https://link.cnki.net/urlid/46.1068.S.20230818.1326.002. (accessed on 26 November 2024). (In Chinese)
14. Jia, Y.Y.; Qiu, Y.P.; Zhou, X.Y.; Geng, Y.H.; Sun, H.N.; Liu, D.Y. Genetic Diversity Analysis of *Clematis macropetala* Based on Morphological Markers and ISSR Molecular Makers. *Acta Agrestia Sin.* **2024**, *32*, 75–86. (In Chinese)
15. Jia, Y.Y.; Zhou, X.Y.; Geng, Y.H.; Sun, H.N.; Liu, D.Y. Identification and genetic diversity analysis of hybrid progeny of *Clematis macropetala* using ISSR markers. *J. Gansu Agric. Univ.* **2024**, *59*, 155–164. (In Chinese)
16. Li, T.; Guo, J.; Li, Y.Y.; Ning, H.; Sun, X.; Zheng, C.S. Genetic diversity assessment of chrysanthemum germplasm using conserved DNA-derived polymorphism markers. *Sci. Hortic.* **2013**, *162*, 271–277. [CrossRef]
17. George, S.; Sharma, J.; Yadon, V.L. Genetic diversity of the endangered and narrow endemic *Piperia yadonii* (Orchidaceae) assessed with ISSR polymorphisms. *Am. J. Bot.* **2009**, *96*, 2022–2030. [CrossRef]
18. Hua, X.F.; Zhong, F.L.; Lin, Y.Z.; Chen, W.H.; Chen, X.; Zhu, H.S. Genetic diversity analysis of strawberry accession with ISSR. *Fujian J. Agric. Sci.* **2013**, *28*, 232–236. (In Chinese)
19. Zhang, Y.F.; Wu, Q.Y.; Zhang, S.H.; Gao, F.J.; Fan, Y.S.; Han, J.L.; Liu, H.Y. Genetic background analysis of strawberry hybrids and parents by ISSR molecular markers. *Southwest China J. Agric. Sci.* **2022**, *35*, 162–167. (In Chinese)
20. Liu, L.L.; Liu, R.M.; Liu, Y.; Li, J.H. Genetic relationship analysis of 12 *Cerasus* spp. germplasms based on SRAP and ISSR marker. *Mol. Plant Breed.* **2024**, *22*, 3247–3252. (In Chinese)
21. Ahmad, F.K.; Noori, I.M. Evaluation of genetic diversity of figs (*Ficus carica* L.) in Sulaymaniyah governorate using morphological, pomological and ISSR molecular marker. *Tikrit J. Agric. Sci.* **2023**, *23*, 147–175. [CrossRef]
22. Collard, B.C.Y.; Mackill, D.J. Conserved DNA-derived polymorphism (CDDP): A simple and novel method for generating DNA markers in plants. *Plant Mol. Biol. Rep.* **2009**, *27*, 558–562. [CrossRef]
23. Luo, S.; Li, B.J.; Fu, Y.Q.; Kong, C.; Yng, J.W.; Gao, X.; Zhao, T.T.; Wang, H.R.; Zhang, Y.L.; Shi, Q.Q. Genetic diversity analysis of basal blotchintree peony petal based on phenotypic clustering and CDDP molecular markers. *J. Northwest A F Univ.* **2024**, *52*, 110. (In Chinese)

24. Jiang, L.Y.; Zang, D.K. Analysis of genetic relationships in *Rosa rugosa* using conserved DNA-derived polymorphism markers. *Biotechnol. Biotechnol. Equip.* **2018**, *32*, 88–94. (In Chinese) [CrossRef]
25. Meriem, A.; Karim, G.; Donia, A.; Marwa, L.; Khaled, C.; Ghada, B.; Amel, S.H. Conserved DNA-derived polymorphism, new markers for genetic diversity analysis of Tunisian *Pistacia vera* L. *Physiol. Mol. Biol. Plants* **2019**, *25*, 1211–1223.
26. Igwe, D.O.; Onyinye, C.I.; Anne, A.O.; George, A.; Ude, G.N. Genetic diversity and population assessment of *Musa* L. (Musaceae) employing CDDP markers. *Plant Mol. Biol. Rep.* **2021**, *39*, 801–820. [CrossRef]
27. Kamaran, S.R.; Hoshman, O.M.; Jamal, M.F.; Djshwar, D.L.; Nawroz, A.T. Genetic diversity and relationships among Iris aucheri genotypes determined via ISSR and CDDP markers. *Genet. Resour. Crop Evol.* **2024**. [CrossRef]
28. Lanja, H.K.; Nawroz, A.T.; Rupak, T.A. Molecular variation in some taxa of genus *Astragalus* L. (Fabaceae) in the Iraqi Kurdistan region. *Horticulturae* **2023**, *9*, 1110. [CrossRef]
29. Zhao, M.Z. *Descriptors and Data Standard for Strawberry (Fragaria spp.)*, 1st ed.; China Agriculture Press: Beijing, China, 2006; pp. 12–35. (In Chinese)
30. Wang, X.W.; Fan, H.M.; Li, Y.Y.; Sun, X. Analysis of genetic relationships in tree peony of different colors using Conserved DNA-derived polymorphism markers. In Proceedings of the 2014 Academic Annual Meeting of Ornamental Horticulture Specialized Committee of Chinese Horticultural Society, Qingdao, China, 16 July 2014; pp. 65–73. (In Chinese)
31. Zhang, Y.; Suo, Y.J.; Sun, P.; Han, W.J.; Diao, S.F.; Li, H.W.; Zhang, J.J.; Fu, J.M.; Li, F.D. Analysis on fruit morphological diversity of persimmon germplasm resources. *Acta Hortic. Sin.* **2022**, *49*, 1473–1490. (In Chinese)
32. David, O.I.; Celestine, A.A.; Benjamin, E.U.; Kenneth, I.O.; Omena, B.O.; George, N.U. Assessment of genetic diversity in *Vigna unguiculata* L. (Walp) accessions using inter-simple sequence repeat (ISSR) and start codon targeted (SCOT) polymorphic markers. *BMC Genet.* **2017**, *18*, 98.
33. Visalakshi, M.; Muthulakshmi, R.; Ganga, M.; Boopathi, N.M.; Vasanth, S.; Ganesh, S. Assessment of genetic diversity among the elite Rose (*Rosa* spp.) accessions using RAPD markers. *Int. J. Environ. Clim. Change* **2023**, *13*, 1305–1311. [CrossRef]
34. Guillaume, E.; Sebastien, R.; Jerome, G. Detecting the number of clusters of individuals using the software STRUCTURE: A simulation study. *Mol. Ecol.* **2005**, *14*, 2611–2620.
35. Liu, L.; Wang, L.; Zhao, C.; Yao, J.; Zhang, F.; Zhang, H.; Ye, Z.; Qin, Z.; Zheng, Y. Genetic Diversity and Alterations of Population Structure in Restorers of Dual Cross-line Hybrid Wheat with Thermo-photoperiod Sensitive Male Sterile. *Chin. J. Biochem. Mol. Biol.* **2009**, *25*, 867–875. (In Chinese)
36. Bai, Y.X.; Zheng, X.Q.; Yao, Y.H.; Yao, X.H.; Wu, K.L. Genetic diversity analysis and comprehensive evaluation of phenotypic traits in Hulless barley germplasm resources. *Sci. Agric. Sin.* **2019**, *52*, 4201–4214. (In Chinese)
37. Liao, G.; Xu, X.; Huang, C.; Zhong, M.; Jia, D.F. Resource evaluation and novel germplasm mining of *Actinidia eriantha*. *Sci. Hortic.* **2021**, *11*, 110037. [CrossRef]
38. Liu, J.Q.; Yin, M.Y.; Zuo, S.Y.; Yang, S.B.; Wu, Y.T.N. Phenotypic variations in natural populations of *Amygdalus pedunculata*. *Chin. J. Plant Ecol.* **2017**, *41*, 1091–1102. (In Chinese)
39. Lu, P.C.; Zheng, Y.; Zhou, X.Q.; Xv, Z.; Ai, Y.; Zhou, Y.Z.; Zhu, W.Y.; Peng, D.H. Phenotypic diversity of 45 cultivars of *Cymbidium tortisepalum*. *Chin. J Tropi. Cro.* **2021**, *42*, 2518–2525. (In Chinese) [CrossRef]
40. Dong, S.J.; Wang, R.X.; Zhang, H.K.; Chen, J.H.; Liu, L.X.; Yu, Q.F. Analysis on diversity of fruit phenotypic characters of *Armeniaca mandshurica* from different provenances. *J. Plant Resour. Environ.* **2020**, *29*, 42–50. (In Chinese)
41. Wu, G.S.; Sun, L.D.; Hao, R.J.; Shi, W.F.; Zhang, J.; Chen, J.X. Study on the phenotypic diversity of *P. mume* Sieb. et Zucc. germplasm resources. *J. Anhui Agric. Sci.* **2011**, *39*, 12008–12009. (In Chinese)
42. Du, K.; Huang, W.Y.; Hou, R.H.; Gao, J.H.; Tang, F. Genetic diversity analysis of *Medicago ruthenica* L. based on phenotypic traits. *Acta Agrestia Sin.* **2024**, 1–20. (In Chinese) [CrossRef]
43. Yang, J.; Zhang, X.M.; Chen, Y.; Fan, D.C.; Wan, T. Genetic diversity analysis of phenotypic traits of 30 *agropyron* genus germplasm resources in Inner mongolia. *Chin. J. Gras.* **2023**, *45*, 1–11. (In Chinese) [CrossRef]
44. Li, Q.Y.; Su, X.J.; Ma, H.H.; Du, K.; Yang, M.; Chen, B.L.; Fu, S.; Fu, T.J.; Xiang, C.; Zhao, Q. Development of genic SSR marker resources from RNA-seq data in *Camellia* japonica and their application in the genus *Camellia*. *Sci. Rep.* **2021**, *11*, 9919. [CrossRef] [PubMed]
45. Yang, Y.S.; Tang, J.M.; Zou, R.; Luo, Y.J.; Deng, Z.H.; Li, D.X.; Chai, S.F.; Wei, X. Study on the genetic diversity of germplasm resources of characteristic medicinal orchid plant *Habenaria dentata*. *Genes* **2023**, *14*, 1749. [CrossRef] [PubMed]
46. Hilde, N. Comparison of different nuclear DNA markers for estimating intraspecific genetic diversity in plants. *Mol. Ecol.* **2004**, *13*, 1143–1155.
47. Yang, Y.; Li, Y.J.; Li, L.; Wu, Y.; Yang, L.; Tian, Y.; Tang, J.; Tang, Y.X. Phenotypic traits and ISSR genetic diversity analysis of poplar clones. *J. Cent. South Univ. For. Technol.* **2024**, *44*, 138–147. (In Chinese)
48. Zhai, C.J.; Ge, L.J.; Cheng, Y.J.; Chou, L.; Wang, X.Q.; Liu, S.D. Genetic diversity analysis of Wax Gourd and Chieh-Qua germplasm resources based on phenotypic traits and SSR markers. *Chin. Agric. Sci.* **2024**, *57*, 3440–3467. (In Chinese)
49. Noguchi, Y.; Morishita, M.; Muro, T.; Kojima, A.; Sakata, Y.; Yamada, T.; Sugiyama, K. 'Tokun': A new aromatic decaploid interspecific hybrid strawberry. *J. Jpn. Assoc. Odor Environ.* **2011**, *42*, 122–128.
50. Wang, Q.L.; Zhao, M.Z.; Wang, Z.W.; Wu, W.M.; Qian, Y.M. 'Zijinhong' a new red-flowered strawberry cultivar. *Acta Hortic. Sin.* **2017**, *44*, 2425–2426. (In Chinese)

51. Chang, L.L.; Dong, J.; Zhong, C.F.; Sun, J.; Sun, R.; Shi, K.; Wang, G.X.; Zhang, Y.T. Pedigree analysis of strawberry cultivars released in China. *J. Fruit Sci.* **2018**, *35*, 158–167. (In Chinese)
52. Nawroz, T.; Djshwar, L.; Kamaran, R.; Didar, R.; Kamil, M.; Shokhan, S.; Avin, M.; Rebwar, A. Assessment of genetic variation and population structure in Iraqi barley accessions using ISSR, CDDP, and SCoT markers. *Czech J. Genet. Plant Breed.* **2023**, *59*, 148–159.

horticulturae

MDPI

Article

Flowering Responses to Vernalization and Photoperiod in *Minuartia laricina* (L.) Mattf., a Perennial Herb in the Korean Peninsula

Dong Gyu Lee [1,†], Suejin Park [2,3,†], Chae Won Kim [1], Jae Hui Kim [1] and Seung Youn Lee [1,4,*]

1 Department of Horticulture and Breeding, Graduate School of Andong National University, Andong 36792, Republic of Korea; dgdg10256@gmail.com (D.G.L.); chaewk13@gmail.com (C.W.K.); wogml0157@naver.com (J.H.K.)
2 Department of Horticulture, Jeonbuk National University, Jeonju 54896, Republic of Korea; sjpark19@jbnu.ac.kr
3 Institute of Agriculture Science & Technology, Jeonbuk National University, Jeonju 54896, Republic of Korea
4 Department of Smart Horticultural Science, Andong National University, Andong 36729, Republic of Korea
* Correspondence: mrbig99@anu.ac.kr; Tel.: +82-54-820-5472
† These authors contributed equally to this work.

Abstract: This study investigated the flowering physiology of *Minuartia laricina* (L.) Mattf. to provide fundamental information for the development of flowering control techniques. A phenology test was conducted in an open field, where *M. laricina* overwintered from December to February. Flower buds began to form in late March as temperatures and photoperiods increased. Flowering started in late April and continued until mid-June, with seeds maturing in capsule-shaped fruits by mid-August. In a greenhouse experiment, *M. laricina* was subjected to cold treatments at 5 °C for 0, 2, 4, 8, or 12 weeks, followed by photoperiod treatments of 9/15, 11/13, 13/11, 16/8, 24/0 h day/night photoperiod, or night interruption (9 h + 4 h). Plant growth and flowering were measured during the photoperiod treatments. The results showed 100% flowering with 8 weeks of cold treatment and under 16 h or longer photoperiod conditions. However, without cold treatments flowering was not observed across all photoperiods. These findings align with outdoor conditions, where *M. laricina* flowers experience a cold winter and under long photoperiods. Thus, *M. laricina* is classified as an obligate vernalization and long-day plant, requiring at least eight weeks of cold and a 16/8 h photoperiod to flower.

Keywords: native plant; obligate vernalization plant; obligate long-day plant; phenology; plant growth

1. Introduction

Native plants have gained attention as essential components in sustainable landscaping and biodiversity conservation. They are highly adaptable near their natural habitats, making them effective replacements for invasive species [1]. Additionally, native plants play a crucial role in ecosystems by supporting wildlife diversity, specifically by providing pollinators with food and habitat resources [2,3]. Recent studies indicated that consumers prefer landscapes featuring a higher proportion of native plants with aesthetic and functional values [4]. Therefore, interest in cultivating native plants has been increasing not only for their ecological roles but also for their ornamental and sustainable values [5]. However, the commercial use of native plants remains limited due to insufficient information on their growth characteristics and the lack of effective cultivation techniques [6,7].

Minuartia laricina (L.) Mattf., a perennial herbaceous plant of the family Caryophyllaceae, is native to Mt. Baekdu, and it is distributed across northern regions of the Korean Peninsula, as well as in Russia, Mongolia, and China. Normally, this species produces 4–5 white flowers arranged in cyme-type inflorescence at the top of each stem [8]. With a compact growth habit and a height of 10–30 cm, *M. laricina* shows considerable potential as a ground cover plant. Recognizing its ornamental potential, the Korean National Arboretum designated *M. laricina* as a new ornamental plant in 2019. Despite its promising traits, research on its flowering characteristics remains insufficient, limiting its application in horticulture practices.

Understanding flowering characteristics is significant not only for the development of cultivation methods, but also for maximizing the ecological benefits of native plants in garden ecosystems. Flowering traits directly influence interactions between flowers and pollinators, thus playing a fundamental role in maintaining ecological balance [9]. Furthermore, elucidating flowering characteristics of native plants provides essential insights for optimizing cultivation techniques and developing ecological guidelines for pollinator-friendly gardens.

Among the factors influencing flowering, vernalization and photoperiod are particularly significant. Vernalization refers to the promotion of flowering through exposure to low temperature ranging 1–7 °C for a specific duration [10,11]. Photoperiod, the length of daylight to which plants are exposed, plays a crucial role in regulating flowering in many plant species [12]. By utilizing the flowering physiological responses of plants, identifying specific environmental conditions that induce flowering will provide key information for the improvement of cultivation techniques and the development of new cultivars.

Previous research on *M. laricina* reported growth and flowering regulation using plant growth retardants [13]. However, the effects of vernalization and photoperiod on the flowering of this species have not been investigated. Therefore, the objective of this study was to elucidate the flowering responses of *M. laricina* to cold treatment and photoperiod. By providing fundamental data on flowering control, this research aims to support the cultivation and industrialization of *M. laricina*, ultimately contributing to its wider use as an ornamental plant in sustainable landscaping.

2. Materials and Methods

2.1. Plants Materials

Seeds of *M. laricina* were sown in June 2022 at a wildflower nursery (Samhan Plants, Yongin-si, Republic of Korea) and subsequently grown in 8 × 7 cm (diameter × height) plastic pots filled with a soilless media (Baroker; Seoul Bio Co., Ltd., Eumseong, Republic of Korea). They were moved to either open field plots or a greenhouse at Andong National University in October 2022 for this study. The plants used in the experiments had an average height of 4.5 ± 0.3 cm and an average width of 7.8 ± 0.6 cm (n = 12).

2.2. Expt. 1: Phenological Observation of Flowering Characteristics in M. laricina

M. laricina was planted in open field plots on 20 October 2022, at Andong National University. Measurements of growth and development were conducted monthly from 20 October 2022 to 28 September 2023, to evaluate flowering characteristics of *M. laricina* in an open field. Plant height and width were measured using a standard ruler, and the number of branches was counted manually. Branch length was determined by measuring the longest branch of each plant. The number of visible buds was recorded, and visible bud rate was calculated as the proportion of plants with at least five visible buds out of 12 plants. Similarly, the number of flowers and flowering rate were recorded. Air temperature was recorded every 10 min using a data logger (WatchDog 1650 Micro Station, Spectrum Technologies, Aurora, IL, USA) (Figure 1).

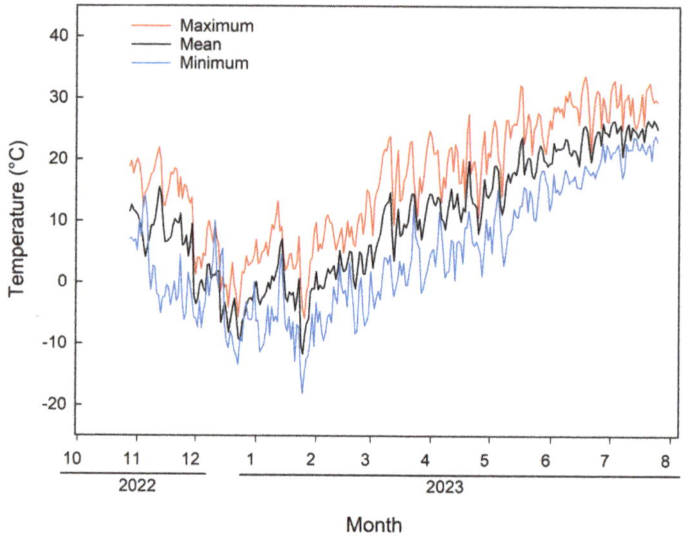

Figure 1. Daily air temperatures in the open field during Expt. 1.

2.3. Expt. 2: Growth and Flowering Response to Cold Treatment and Photoperiod

M. laricina plants grown in 8 × 7 cm plastic pots were subjected to cold treatments at 5 °C in a cold storage chamber for 0, 2, 4, 8, or 12 weeks starting from 20 October 2022. During the cold treatment, plants were placed under a light intensity of 3–4 $\mu mol \cdot m^{-2} \cdot s^{-1}$ provided by a fluorescent lamp (30W FQU30EXN-D; Shanghai Haneup City Lighting Electric Co., Ltd., Shanghai, China) for 9 h per day. Plants were irrigated with tap water twice a week to prevent the potting media from drying.

Following each period of cold treatment, plants were transplanted into 12 cm × 10.5 cm (diameter × height) plastic pots filled with horticultural substrate (Sunshine Mix #4; Sun-Gro Horticulture, Agawam, MA, USA). Subsequently, 3 g of a controlled-release fertilizer (PurKote 14-14-14 [90-day]; Pursell, Sylacauga, AL, USA) was applied to each pot, and the plants were then subjected to various photoperiods.

The photoperiod treatments included 9/15, 11/13, 13/11, 16/8, and 24/0 h day/night photoperiods, as well as a night interruption (NI) treatment. In this study, we adopted these specific photoperiods of 9, 11, 13, 16 h, and continuous light (24 h) based on the commonly used methodology in horticultural and plant physiology research that centers around a 24 h cycle [14]. For these treatments, plants were exposed to natural light from 9:00 a.m. to 6:00 p.m., providing a 9 h day length. To extend the photoperiod, two fluorescent lamps (30 W FQU30EXN-D; Shanghai Haneup City Lighting Electric Co., Ltd., Shanghai, China) were used per bench, and black curtains were drawn around each bench to block external light and reduce the influence between photoperiod treatments. The light intensity during the photoperiod extension was maintained at 3–4 $\mu mol \cdot m^{-2} \cdot s^{-1}$ to ensure consistency of daily light integral (DLI). In the NI treatment, the photoperiod was extended from 10:00 p.m. to 2:00 a.m. the following day. It is well documented that under short-day conditions, interrupting the dark period with supplemental light can stimulate flowering in long-day plants (LDP) while concurrently suppressing flowering in short-day plants (SDP). In general, a 4 h NI photoperiod (from 22:00 to 02:00) is recommended to promote flowering in LDP, and this approach has been shown to be effective to induce flowering in various herbaceous perennial species [15,16]. During all photoperiod treatments, tap water was applied by overhead watering every three days.

Photoperiod treatments were conducted for 15 weeks in a greenhouse at Andong National University, where environmental conditions were continuously monitored using data loggers (WatchDog 1650 Micro Station, Spectrum Technologies, IL, USA). The air temperature and daily light integral (DLI) in the greenhouse as well as for the benches of each photoperiod treatment are shown in Figure 2 and Table 1, respectively.

Table 1. Air temperature and daily light integral for the benches of each photoperiod treatment in Expt. 2.

Cold Treatment (Weeks)	Photoperiod	Avg. Daily Air Temperature (°C)		DLI ($mol \cdot m^{-2} \cdot d^{-1}$)	
0	9 h	16.0	±0.9 [1]	7.2	±0.3 [1]
	11 h	15.7	±1.0	6.8	±0.3
	13 h	15.9	±1.0	6.1	±0.2
	16 h	16.5	±1.1	5.9	±0.2
	24 h	17.9	±1.0	6.8	±0.3
	NI [2]	17.8	±0.9	7.0	±0.3
2	9 h	15.9	±0.9	7.0	±0.3
	11 h	15.5	±0.9	6.5	±0.3
	13 h	15.7	±1.0	5.9	±0.2
	16 h	16.3	±1.0	5.8	±0.2
	24 h	17.6	±0.9	6.5	±0.3
	NI	17.7	±0.9	6.9	±0.3
4	9 h	16.1	±1.0	7.7	±0.4
	11 h	15.7	±1.1	7.2	±0.4
	13 h	15.9	±1.1	6.6	±0.3
	16 h	16.5	±1.0	6.6	±0.3
	24 h	17.6	±0.9	7.3	±0.4
	NI	17.9	±0.9	7.6	±0.4
8	9 h	17.3	±1.1	7.9	±0.4
	11 h	17.3	±1.2	8.1	±0.4
	13 h	17.3	±1.1	8.0	±0.4
	16 h	17.6	±1.1	8.2	±0.4
	24 h	18.6	±1.0	8.1	±0.4
	NI	18.7	±0.8	7.9	±0.4
12	9 h	20.5	±1.3	7.8	±0.4
	11 h	20.5	±1.4	8.2	±0.4
	13 h	20.7	±1.3	9.0	±0.4
	16 h	20.8	±1.2	9.5	±8.1
	24 h	21.1	±1.3	8.1	±0.4
	NI	20.6	±1.2	8.0	±0.4

[1] Means ± SD of average daily air temperature and daily light integral (DLI). [2] NI means 9 h with a 4 h night interruption (22:00–02:00 HR).

During 15 weeks of photoperiod treatments, plant growth and development were measured as described above. Additionally, the time to flowering was calculated as the number of days from the formation of at least five visible buds to the blooming of at least five flowers. Measurements were conducted weekly until at least five visible buds formed, after which they were taken every three days.

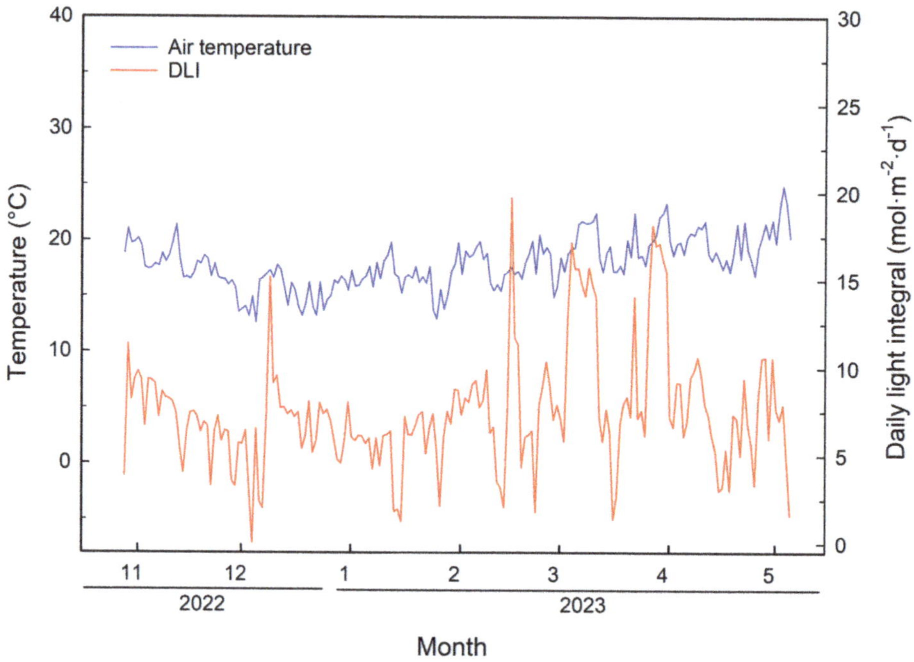

Figure 2. Average air temperature and daily light integral in the greenhouse during Expt. 2.

2.4. Experimental Design and Statistical Analysis

Expt. 1. was conducted using a completely randomized design (CRD) with 12 replicates (n = 12), while Expt. 2. was laid in a 5 × 6 factorial arrangement (cold treatment × photoperiod) in a CRD with 12 replicates (n = 12). Each plant was analyzed as one replicate. Analysis of variance was performed using SAS 9.4 (SAS Institute Inc., Cary, NC, USA), and differences among the treatment means were assessed by Duncan's multiple range test ($p < 0.05$). Regression analysis was conducted to evaluate the linear relationship between the number of flowers and the duration of cold treatment.

3. Results

3.1. Expt. 1: Phenological Observation of Flowering Characteristics in M. laricina

This experiment was conducted to investigate the growth and flowering characteristics of *M. laricina* under outdoor garden conditions. From November to January, a period with relatively low temperatures, *M. laricina* exhibited minimal growth. However, as temperatures and day lengths increased from March to August, plant height, width, number of branches, and branch length significantly increased (Figure 3A,B). Visible buds were first observed in late March, and flowering was initiated in late April, consistently producing over 80 flowers each until mid-June (Figures 3C and 4D,E). Visible buds and flowers were observed in all tested plants between May and July (Figure 3D). *M. laricina* plants started producing capsule-shaped fruits in late June and their leaves and branches turned to brown in late September (Figure 4F).

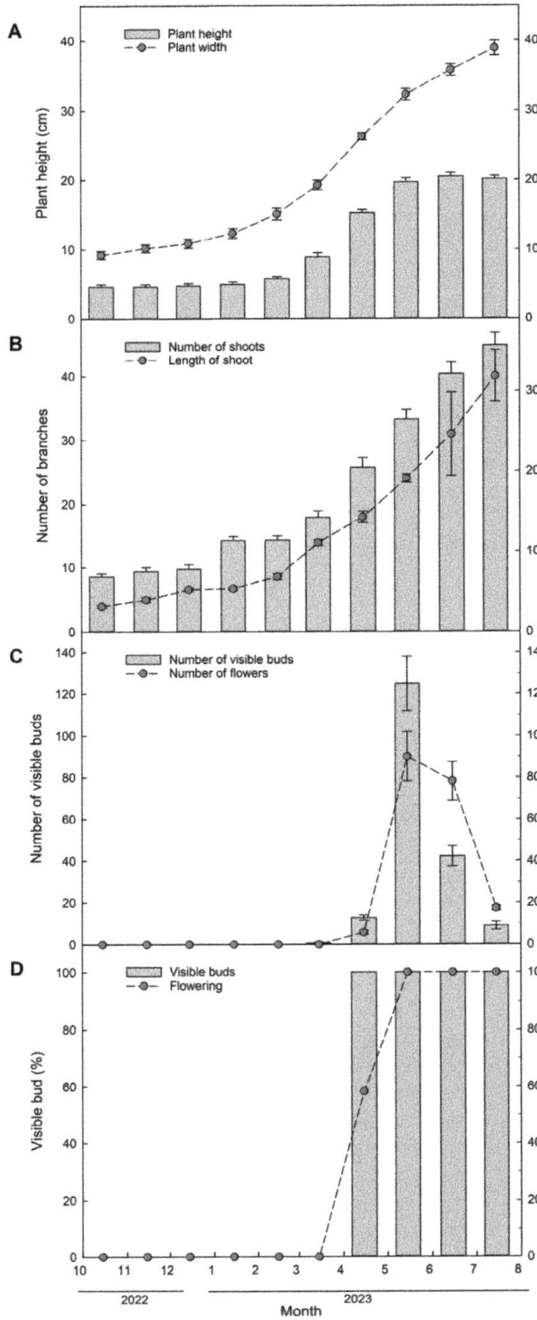

Figure 3. Seasonal changes in growth characteristics of *Minuartia laricina* (L.) Mattf. in the open field experiment. Plant height and width (**A**), the number of branches and branch length (**B**), the number of visible buds and flowers (**C**), and visible bud and flowering rate (**D**) were assessed. Vertical bars are standard errors.

Figure 4. Seasonal changes in the phenotype of *Minuartia laricina* (L.) Mattf. during Expt. 1. on 20 October 2022 (**A**), 21 January 2023 (**B**), 21 March 2023 (**C**), 12 April 2023 (**D**), 10 May 2023 (**E**), and 20 September 2023 (**F**).

3.2. Expt. 2: Growth and Flowering Response to Cold Treatment and Photoperiod

This experiment investigated the effects of the duration of cold treatment and photoperiod. The results showed that cold treatment significantly influences plant height, width, the number of branches, and flower production (Table 2). Plant height increased with longer cold treatment durations, peaking at 12.5 cm under 12 weeks of cold treatment. Plant width and branch number also increased with extended cold treatment, with the widest plants (32.7 cm) and the most branches (38.6) recorded under 12 weeks of cold treatment and 24 h photoperiod. Both photoperiod and its interaction with cold treatment had significant effects, with longer photoperiods (16 or 24 h) generally promoting better growth and development.

Without cold treatment, no flowering was observed across photoperiods (Table 2 and Figures 5–7). While at least four weeks of cold treatment induced flowering, with the highest flower number (74.5) observed under 12 weeks of cold treatment combined with 24 h photoperiod. Twelve-week cold treatment resulted in flowering rates of 16%, 67%, 100%, 100%, 100%, and 100% under 9 h, 11 h, 13 h, 16 h, 24 h photoperiods, and NI, respectively, (Figure 5A). As the duration of cold treatment and day length increased, the flowering rate also increased. Moreover, the number of days until flowering decreased with prolonged cold treatment, reaching the shortest time (56.2 days) under 12-week cold treatment and 24 h photoperiod (Figure 5B). Flowering was the fastest under 24 h photoperiod, followed by NI, 16 h, 13 h, and 11 h photoperiods, respectively, after at least eight weeks of cold treatment.

Regression analysis revealed that both cold treatment duration and photoperiod significantly influence the flower number of *M. laricina* (Figure 7). The number of flowers increased as the duration of cold treatment extended across all photoperiod conditions, with the highest regression coefficients observed under 24 h (6.8599 ***) and NI (3.0930 ***) photoperiods, showing the strongest correlations (R^2 = 0.7282 and 0.6041, respectively). Conversely, the flower number under the 9 h photoperiod showed the lowest coefficient (0.1333 *) and correlation (R^2 = 0.0906), indicating minimal response to cold treatment. At least eight weeks of cold treatment were necessary to achieve substantial flowering, with longer photoperiods (16 h, 24 h, and NI) increasing flower production. The regression equations and R^2 values indicate that cold treatment duration strongly promoted flowering under extended photoperiods, with the effect being less pronounced under shorter photoperiods (9 h and 11 h). These results highlight that longer cold treatment durations, combined with extended photoperiods, are critical for maximizing flowering in *M. laricina*.

Table 2. Growth characteristics of *Minuartia laricina* (L.) Mattf at 15 weeks after photoperiod following cold treatment.

Cold Treatment (Weeks)	Photoperiod	Plant Height (cm)		Plant Width (cm)		Number of Branches		Number of Flowers	
0	9 h	3.1	b [1]	11.1	c	15.6	d	0	
	11 h	3.0	b	11.2	c	17.4	c	0	
	13 h	3.2	b	12.5	c	18.4	c	0	
	16 h	3.6	b	14.5	b	21.3	b	0	
	24 h	5.9	a	23.5	a	24.3	a	0.3	
	NI	3.5	b	15.2	b	20.8	b	0	
2	9 h	3.5	bc	12.0	d	16.7	d	0	b
	11 h	3.6	bc	13.5	d	17.7	cd	0	b
	13 h	3.0	c	15.9	c	18.7	c	0	b
	16 h	3.8	bc	21.2	b	21.3	b	0.5	b
	24 h	5.4	a	25.9	a	24.8	a	9.3	a
	NI	4.0	b	20.8	b	21.3	b	0	b
4	9 h	4.0	c	16.4	d	16.4	e	0	c
	11 h	4.2	c	18.8	c	18.6	de	0.2	c
	13 h	4.2	c	19.9	c	19.3	cd	1.3	c
	16 h	5.8	ab	22.0	b	21.5	bc	8.9	b
	24 h	6.8	a	26.1	a	24.0	a	18.8	a
	NI	5.2	bc	22.8	b	21.8	ab	12.4	b
8	9 h	4.0	d	19.2	d	17.8	c	0.2	c
	11 h	4.8	cd	20.9	d	18.0	c	0.9	c
	13 h	6.0	bc	25.8	c	19.2	c	11	c
	16 h	7.7	a	29.8	b	27.0	b	31.7	b
	24 h	8.5	a	32.3	a	33.5	a	65.8	a
	NI	7.3	ab	29.5	b	28.8	b	29.9	b
12	9 h	4.8	e	18.1	d	18.8	c	1.8	c
	11 h	7.2	d	22.2	c	18.8	c	7.8	c
	13 h	9.2	c	27.1	b	20.3	c	23.6	b
	16 h	10.8	b	29.0	b	28.8	b	29.3	b
	24 h	12.5	a	32.7	a	38.6	a	74.5	a
	NI	10.6	bc	28.9	B	29.6	b	32	b
Significance									
Cold treatment (C)		***		***		***		***	
Photoperiod (P)		***		***		***		***	
C × P		***		***		***		***	

[1] Different letters within each cold treatment indicate significant differences at $p < 0.05$ (Duncan's multiple range tests). *** means significant at $p < 0.001$, respectively.

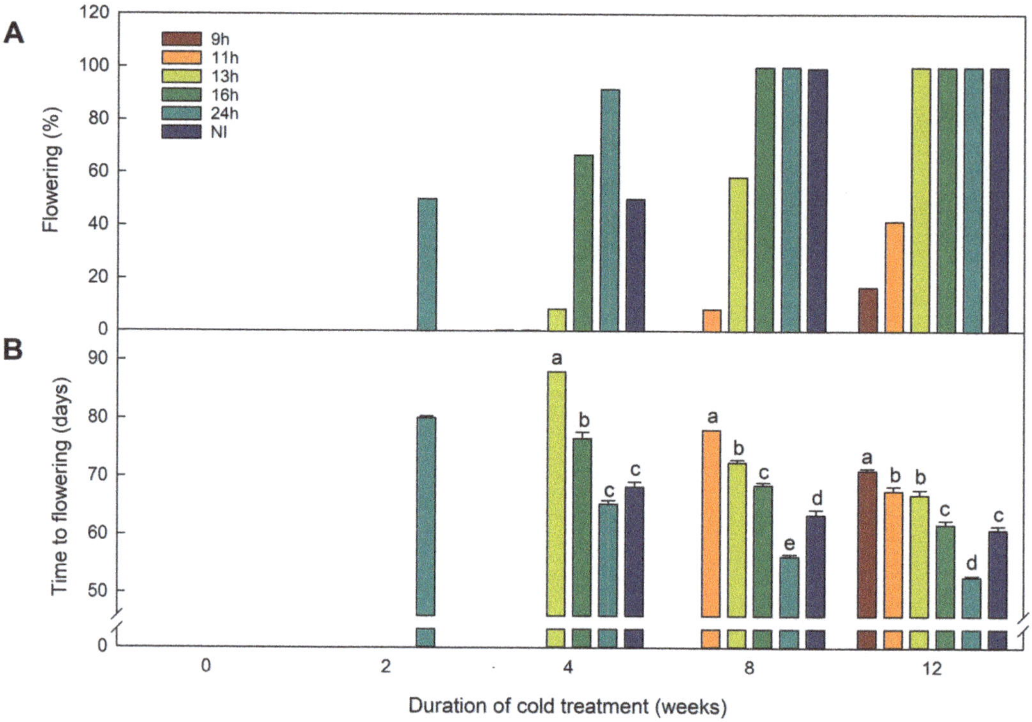

Figure 5. Effects of cold duration and photoperiods on flowering rate (**A**) and the time to flowering (**B**) in *Minuartia laricina* (L.) Mattf. Different letters within each cold treatment indicate significant differences at $p < 0.05$ (Duncan's multiple range tests). Vertical bars represent standard errors.

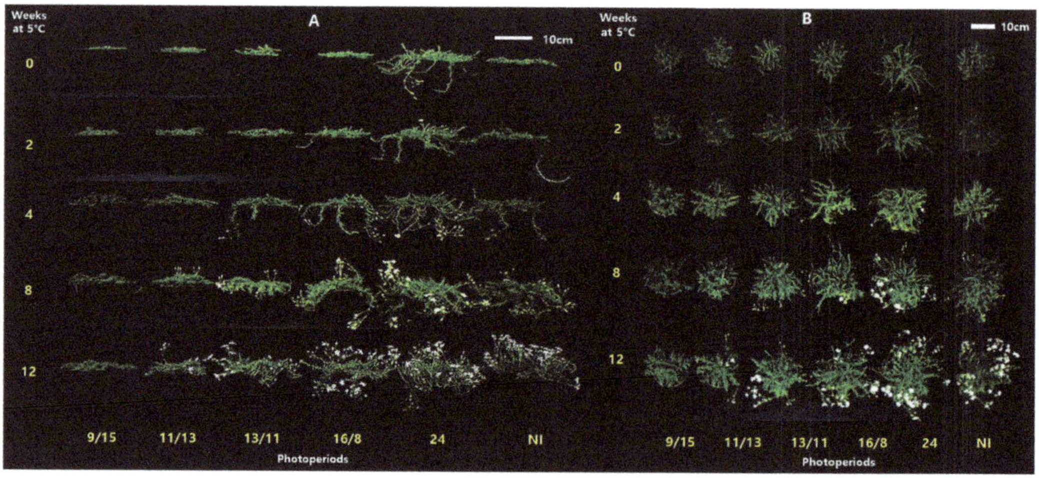

Figure 6. Effects of cold treatments and photoperiods on growth and flowering of *Minuartia laricina* (L.) Mattf. The photos were taken at 15 weeks after photoperiod following cold treatment: front view (**A**), and top view (**B**).

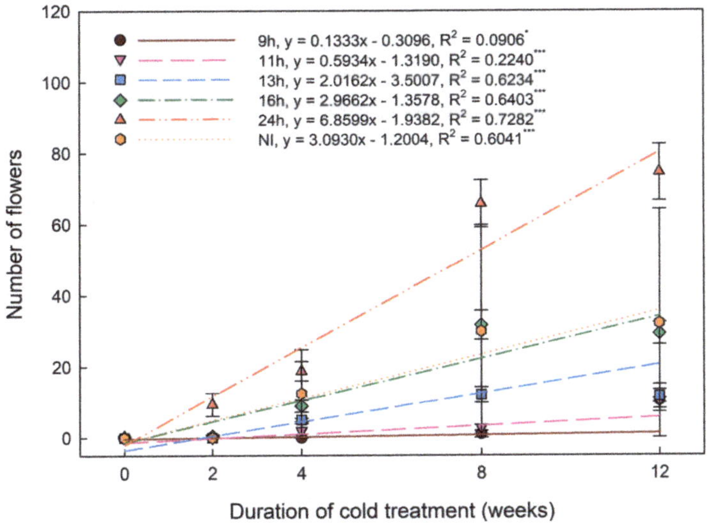

Figure 7. Relationship between the duration of cold treatment and the number of flowers in *Minuartia laricina* (L.) Mattf. under various photoperiods at 15 weeks after cold treatment. The photoperiod treatments are 9/15, 11/13, 13/11, 16/8, 24/10 (day/night) h, and night interruption (NI). Each data point represents the average of flower numbers, but the regression analysis was performed with all data. Bars represent standard errors. *, significant at $p < 0.05$; ***, significant at $p < 0.001$.

4. Discussion

Phenology refers to the recurring seasonal cycles of biological events in organisms, particularly in response to climatic variations, providing a model for understanding plant responses to these changes [17]. Therefore, the phenology experiment allowed us to confirm that *M. laricina* flowers in spring as temperatures increase and photoperiods become longer after a cold winter. According to World Flora Online data, *M. laricina* is known to flower from July to September [18]. However, the results of this experiment indicated that the flowering of *M. laricina* occurred from late April to mid-June (Figure 3). A study of *Patrinia rupestris* reported that the flowering period differed from that observed in the natural habitat, attributing this discrepancy to differences between experimental conditions and the natural environment [19]. Northern plant species are highly sensitive to temperature changes during spring and summer, which play crucial roles in regulating growth and flowering [20,21]. Although the Mt. Baekdu region, where *M. laricina* is native, can be classified as a temperate area, it is characterized as both of temperate and cold climate. Therefore, the region where this phenology study was conducted has shorter winters and an earlier onset of temperature increases compared to the natural habitat, resulting in earlier flowering of *M. laricina*.

To identify the optimum conditions for growth and flowering in *M. laricina*, flowering responses were characterized under various cold treatments and photoperiods. The flowering rate increased with longer periods of cold treatment, being particularly high under the 12-week cold treatment condition (Figure 5). Some plants require sufficient cold treatment to break dormancy, which serves as a crucial survival strategy to prevent damage from sudden temperature drops following spring budburst [9,22]. The effectiveness of cold treatment in promoting flowering varies in terms of temperature and duration depending on the plant species [23]. Based on their cold requirements for flowering, plants can be broadly classified into three categories: non-vernalization plants, which flower without cold treatment; facultative vernalization plants, for which cold treatment is not essential but promotes flowering; and obligate vernalization plants, which require cold treatment for flowering [24,25]. Studies classifying flowering responses to cold treatment in perennial herbaceous plants have categorized *Veronica kiusiana* var. *diamantiaca* [26], *Lysimachia mauritiana* Lam. [27], *Oenothera fruticose* [28], and *Corydalis speciosa* [29] as obligate vernalization plants, which necessitate cold treatment for flowering. In contrast, *V. rotunda* var. *subintegra* (Nakai) T. Yamaz., *V. longifolia* L. [30], and *Aquilegia* 'Origami Blue and White' [31] have been reported to flower without cold treatment. According to the results of this study, *M. laricina* is considered an obligate vernalization plant as it required at least four weeks of cold conditions for flowering.

Plants that exhibit enhanced flowering under long-day conditions are classified as long-day plants, and this characteristic enables them to recognize seasonal changes and transition to the reproductive growth stage [32]. If the critical day length is not met, the flowering of long-day plants is either delayed or does not occur [33]. Long-day plants can be broadly categorized into obligate long-day plants and facultative long-day plants based on their requirements for day length [22,24,34]. Obligate long-day plants flower only when exposed to day lengths exceeding a specific critical day length, and they do not flower under shorter day lengths. Facultative long-day plants, on the other hand, experience enhanced flowering under certain day lengths, but they do not require a specific critical day length to flower. In other words, while extended day lengths promote flowering, facultative long-day plants can still flower under relatively shorter day lengths. Obligate long-day plants tend to exhibit a significant increase in flowering rate once the critical day length is surpassed. In studies classifying flowering responses to photoperiod in perennial herbaceous plants, *Rudbeckia fulgida* 'Goldsturm' was categorized as an obligate long-day plant. Research investigating flowering responses to cold treatment and photoperiod conditions found that the critical day length before and after cold treatment was approximately 14 h and 13 h, respectively. Although cold treatment shortened the critical day length, flowering only proceeded under specific day lengths exceeding a certain threshold, thus classifying it as an obligate long-day plant [35]. In contrast, *Leucanthemum* × *superbum* 'Snowcap' [36] and *Lobelia* × *speciosa* 'Compliment Scarlet' [37] were classified as obligate long-day plants when not subjected to cold treatment. However, after cold treatment, they were able to flower without requiring a specific critical day length, thus being reclassified as facultative long-day plants. Similarly, the response to day length following cold treatment can vary [23]. In this study, *M. laricina* achieved a 100% flowering rate under photoperiods exceeding 13/11 h following a 12-week cold treatment (Figures 5–7). Conversely, under photoperiods of 13/11 h or less, the flowering rate was very low, and flowering was considered unsuccessful (Figures 5–7). Some individuals that did flower under relatively shorter photoperiods were likely influenced by complex factors within the greenhouse cultivation environment. Therefore, *M. laricina* was classified as an obligate long-day plant, requiring specific critical day lengths for flowering.

Cold treatment and long-day conditions influence not only flowering but also the overall plant growth. Cold treatment has been reported to promote the production of growth hormones such as gibberellin, thereby inducing plant elongation [33]. Additionally, an increase in the number of branches leads to enhanced photosynthetic capacity, thereby promoting overall plant growth and development. In a related study on *Pisum sativum*, cold and long-day conditions were reported to promote branch formation through the interaction of hormones such as auxin and cytokinin [38]. Therefore, the increased plant growth observed under relatively extended periods of cold treatment and long-day conditions is presumed to be attributable to growth-related hormones. Furthermore, considering that analysis of hormones was not conducted, follow-up studies are needed to elucidate the effects of hormonal changes on plant growth.

Comparing these two experiments, it appears that *M. laricina* in Expt. 1 underwent a prolonged period of natural cold exposure during winter, thereby satisfying its vernalization requirement. However, as daily mean temperatures remained below 10 °C in early spring, the transition from vegetative growth to flowering was delayed. By May, when the daily mean temperature was above 10 °C, flowering occurred. In contrast, plants in Expt. 2 received an 8-week cold treatment under controlled conditions and were then immediately exposed to relatively warm greenhouse temperatures. After vernalization, the warmer temperature, referred to as the forcing temperature, facilitated earlier flowering. The forcing temperature is defined as the post-vernalization temperature that stimulates and accelerates plant development, particularly the transition to flowering. The optimal range of forcing temperature may vary not only between different species but also within a single species, reflecting their respective climatic origins [39]. These observations align with findings for *Oenothera fruticosa*, in which increasing the forcing temperature from 14 to 29 °C significantly reduced the time to visible bud formation, shortened the bud-to-flower time, and accelerated the overall stages of flowering [28]. Similarly, although *M. laricina* satisfied its vernalization requirement in both experimental conditions, the higher temperature in Expt. 2 expedited the flowering process compared with the more gradual warming under natural conditions in Expt. 1.

During the phenological observations, several pollinator species were observed visiting *M. laricina* flowers, including *Seladoni aeraria*, *Apis mellifera*, *Andrena* sp., *Eristalis cerealis*, *Pseudozizeeria maha*, and *Formicidae* sp. (Figure 8). These findings highlight the potential of *M. laricina* in supporting biodiversity. Specifically, native plants are known to be more preferred by indigenous pollinators due to their long-term adaptation to the local environment [40]. Furthermore, pollinator gardens utilizing native plants can enhance biodiversity by providing habitats and food sources for various pollinators [41]. If the flowering control and cultivation techniques for *M. laricina* are successfully developed and commercialized for use in garden landscapes, it could contribute to the creation of pollinator gardens and enhance the ecological value of garden environments.

Figure 8. Observed pollinators during Expt. 1 with *Minuartia laricina* (L.) Mattf. *Seladonia aeraria* (**A**), *Apis mellifera* (**B**), *Andrena* sp. (**C**), *Eristalis cerealis* (**D**), *Pseudozizeeria maha* (**E**), and *Formicidae* sp. (**F**).

5. Conclusions

M. laricina exhibited a flowering period from late April to mid-June under outdoor garden conditions. The greenhouse experiment revealed that successful flowering required a cold treatment at 5 °C for a minimum of 8 weeks and photoperiods longer than 16 h. Consequently, *M. laricina* has an obligate requirement for both cold and long-day photoperiods for flowering.

In aspects of commercial production, accurately controlling cold treatment durations and extending photoperiods can be used to schedule flowering to meet market demands, ensuring a more predictable and profitable supply. By manipulating vernalization and daylength, growers can optimize plant growth cycles and schedule their crops to match specific shipping or retail periods. In terms of conservation, replicating natural seasonal cues is essential for achieving normal flowering in seed collection programs, supporting the preservation and potential reintroduction of *M. laricina* or other similarly adapted species into their native habitats.

This study provides important data to develop cultivation techniques for *M. laricina*. The successful cultivation and flowering of *M. laricina* under controlled conditions highlight its potential as a key plant for the garden industry, which increasingly demands a variety of species. Furthermore, *M. laricina* can contribute significantly to sustainable landscaping and biodiversity conservation, supporting the creation of pollinator-friendly environments.

Author Contributions: Conceptualization, D.G.L., S.P. and S.Y.L.; Investigation, D.G.L., C.W.K. and J.H.K.; data curation, D.G.L.; writing—original draft preparation, D.G.L., S.P. and S.Y.L.; writing—review and editing, S.P. and S.Y.L. All authors have read and agreed to the published version of the manuscript.

Funding: This research was funded by the Basic Science Research Program through the National Research Foundation of Korea (NRF) funded by the Ministry of Education (NRF-2018R1A6A1A03024862).

Data Availability Statement: The original contributions presented in this study are included in the article. Further inquiries can be directed to the corresponding author.

Acknowledgments: Thanks to every member of Floriculture & Landscape Plants Lab (FLPL, Andong National University).

Conflicts of Interest: The authors declare no conflicts of interest.

References

1. Ryu, B.Y. Selection of native plants for the use of veranda container garden at apartment house. *J. Korean People Plants Environ.* **2004**, *7*, 19–23.
2. Burghardt, K.T.; Tallamy, D.W.; Shriver, W.G. Impact of native plants on bird and butterfly biodiversity in suburban landscapes. *Conserv. Biol.* **2009**, *23*, 219–224. [CrossRef]
3. Pardee, G.L.; Philpott, S.M. Native plants are the bee's knees: Local and landscape predictors of bee richness and abundance in backyard gardens. *Urban Ecosyst.* **2014**, *17*, 641–659. [CrossRef]
4. Gillis, A.J.; Swim, J.K. Adding native plants to home landscapes: The roles of attitudes, social norms, and situational strength. *J. Environ. Psychol.* **2020**, *72*, 101519. [CrossRef]
5. Rupp, L.A.; Anderson, R.M.; Klett, J.; Love, S.L.; Goodspeed, J.; Gunnell, J. Native and adapted plant introduction for low-water landscaping. *HortTechnology* **2018**, *28*, 431–435. [CrossRef]
6. Fausey, B.A.; Cameron, A.C. Evaluating herbaceous perennial species as new flowering potted crops. *Acta Hortic.* **2005**, *683*, 207–214. [CrossRef]
7. Wilde, H.D.; Gandhi, K.J.; Colson, G. State of the science and challenges of breeding landscape plants with ecological function. *Hortic. Res.* **2015**, *2*, 14054. [CrossRef]
8. National Institute of Biological Resources. National Biodiversity Information Sharing System. Available online: https://species.nibr.go.kr/home/mainHome.do?cont_link=009&subMenu=009002&contCd=009002&pageMode=view&ktsn=120000060765 (accessed on 1 June 2024).
9. Fenster, C.B.; Armbruster, W.S.; Wilson, P.; Dudash, M.R.; Thomson, J.D. Pollination Syndromes and Floral Specialization. *Annu. Rev. Ecol. Evol. Syst.* **2004**, *35*, 375–403. [CrossRef]
10. Amasino, R.M. Vernalization, competence, and the epigenetic memory of winter. *Plant Cell* **2004**, *16*, 2553–2559. [CrossRef]

1. Kim, D.H.; Doyle, M.R.; Sung, S.; Amasino, R.M. Vernalization: Winter and the timing of flowering in plants. *Annu. Rev. Cell Dev. Biol.* **2009**, *25*, 277–299. [CrossRef] [PubMed]
2. Osnato, M.; Cota, I.; Nebhnani, P.; Cereijo, U.; Pelaz, S. Photoperiod control of plant growth: Flowering time genes beyond flowering. *Front. Plant Sci.* **2022**, *12*, 805635. [CrossRef] [PubMed]
3. Park, J.W.; Shin, U.S.; Oh, H.J.; Lee, S.I.; Kwon, Y.E.; Lee, S.Y.; Jeong, M.J. Effect of plant growth retardants on growth and flowering of *Minuartia laricina* (L.) Mattf. *Flower Res. J.* **2022**, *30*, 92–99. [CrossRef]
4. Thomas, B.; Vince-Prue, D. *Photoperiodism in Plants*; Elsevier: Amsterdam, The Netherlands, 1996.
5. Runkle, E.S.; Heins, R.D. Photocontrol of flowering and extension growth in the long-day plant pansy. *J. Am. Soc. Hortic. Sci.* **2003**, *128*, 479–485. [CrossRef]
6. Chung, S.E.; Park, J.W.; Im, N.H.; Lee, H.B. Long-day Requirement for Floral Induction in *Lythrum salicaria* 'Dwarf Pink'. *J. People Plants Environ.* **2024**, *27*, 269–278. [CrossRef]
7. Katal, N.; Rzanny, M.; Mäder, P.; Wäldchen, J. Deep learning in plant phenological research: A systematic literature review. *Front. Plant Sci.* **2022**, *13*, 805738. [CrossRef]
8. The World Flora Online. Available online: https://www.worldfloraonline.org/taxon/wfo-0001292108 (accessed on 1 June 2024).
9. Song, S.J.; Shin, U.S.; Kim, S.Y.; Jeong, M.J.; Lee, S.Y. Growth and flowering response of *Patrinia rupestris* in response to different cold durations and photoperiods. *Flower Res. J.* **2019**, *27*, 101–108. [CrossRef]
10. Xiang, K.; Guo, Q.; Zhang, B.; Wang, J.; Jin, N.; Wang, Z.; Liu, J.; Wang, C.; Du, Z.; Wang, L.; et al. Impact of preseason climate factors on vegetation photosynthetic phenology in mid–high latitudes of the northern hemisphere. *Plants* **2024**, *13*, 1254. [CrossRef]
11. Ettinger, A.K.; Chamberlain, C.J.; Morales-Castilla, I.; Buonaiuto, D.M.; Flynn, D.F.B.; Savas, T.; Wolkovich, E.M. Winter temperatures predominate in spring phenological responses to warming. *Nat. Clim. Chang.* **2020**, *10*, 1137–1142. [CrossRef]
12. Blanchard, M.G.; Runkle, E.S. Intermittent light from a rotating high-pressure sodium lamp promotes flowering of long-day plants and seedlings. *HortScience* **2010**, *45*, 236–241. [CrossRef]
13. Lang, A. Physiology of flower initiation. In *Encyclopedia of Plant Physiology*; Ruhland, W., Ed.; Springer: Berlin, Germany, 1965.
14. Michigan State University Extension Floriculture & Greenhouse Crop Production. Life After Cold! Available online: https://www.canr.msu.edu/resources/vernalization-part-4 (accessed on 1 June 2024).
15. Michigan State University Extension Floriculture & Greenhouse Crop Production. Cool Campanula. Available online: https://www.canr.msu.edu/resources/cool-campanula (accessed on 1 June 2024).
16. Song, S.J.; Jeong, M.J.; Lee, S.Y. Ecophysiology of growth and flowering of four Korean native *Veronica* taxa in response to the photoperiod and cold treatment. *Acta Hortic.* **2020**, *1291*, 205–214. [CrossRef]
17. Im, N.H.; Lim, S.H.; Lee, H.B.; An, S.K.; Lee, S.Y.; Kim, K.S. Growth and flowering responses of *Lysimachia mauritiana* Lam. to cold treatment and photoperiod. *Sci. Hortic.* **2020**, *270*, 109429. [CrossRef]
18. Clough, E.A.; Cameron, A.C.; Heins, R.D.; Carlson, W.H. Growth and development of *Oenothera fruticosa* is influenced by vernalization duration, photoperiod, forcing temperature, and plant growth regulators. *J. Am. Soc. Hortic. Sci.* **2001**, *126*, 269–274. [CrossRef]
19. Park, J.W.; Im, N.H.; Lee, H.B. *Corydalis speciosa* Maxim. is a day-neutral plant with an obligate cold requirement. *Sci. Hortic.* **2024**, *325*, 112686. [CrossRef]
20. Lim, S.H.; Im, N.H.; An, S.K.; Lee, H.B.; Kim, K.S. Korean native *Veronica rotunda* and *Veronica longifolia* are day-neutral plants with no vernalization requirements. *Hortic. Environ. Biotechnol.* **2021**, *62*, 859–869. [CrossRef]
21. Whitman, C.M.; Runkle, E.S. Determining the flowering requirements of two *Aquilegia* cultivars. *HortScience* **2012**, *47*, 1261–1264. [CrossRef]
22. Mattson, N.S.; Erwin, J.E. The impact of photoperiod and irradiance on flowering of several herbaceous ornamentals. *Sci. Hortic.* **2005**, *104*, 275–292. [CrossRef]
23. Taiz, L.; Zeiger, E.; Møller, I.M.; Murphy, A. *Plant Physiology and Development*, 6th ed.; Oxford University Press: Oxford, UK, 2018.
24. Ha, T.M. A review of plants' flowering physiology: The control of floral induction by juvenility, temperature and photoperiod in annual and ornamental crops. *Asian J. Agric. Food Sci.* **2014**, *2*, 186–195.
25. Runkle, E.S.; Heins, R.D.; Cameron, A.C.; Carlson, W.H. Photoperiod and cold treatment regulate flowering of *Rudbeckia fulgida* 'Goldsturm'. *HortScience* **1999**, *34*, 55–58. [CrossRef]
26. Runkle, E.S.; Heins, R.D.; Cameron, A.C.; Carlson, W.H. Flowering of *Leucanthemum* × *superbum* 'Snowcap' in response to photoperiod and cold treatment. *HortScience* **1998**, *33*, 1003–1006. [CrossRef]
27. Runkle, E.S.; Heins, R.D.; Cameron, A.C.; Carlson, W.H. Cold treatment modifies the photoperiodic flowering response of *Lobelia* × *speciosa*. *Sci. Hortic.* **1999**, *80*, 247–258. [CrossRef]
28. Ferguson, B.J.; Beveridge, C.A. Roles for auxin, cytokinin, and strigolactone in regulating shoot branching. *Plant Physiol.* **2009**, *149*, 1929–1944. [CrossRef]

39. Roberts, E.; Summerfield, R. Measurement and Prediction of Flowering in Annual Crops. In *Manipulation of Flowering*; Atherton, J.G., Ed.; Butterworth's: London, UK, 1987; pp. 17–51.
40. Harmon-Threatt, A.N.; Kremen, C. Bumble bees selectively use native and exotic species to maintain nutritional intake across highly variable and invaded local floral resource pools. *Ecol. Entomol.* **2015**, *40*, 471–478. [CrossRef]
41. Menz, M.H.M.; Phillips, R.D.; Winfree, R.; Kremen, C.; Aizen, M.A.; Johnson, S.D.; Dixon, K.W. Reconnecting plants and pollinators: Challenges in the restoration of pollination mutualisms. *Trends Plant Sci.* **2011**, *16*, 4–12. [CrossRef] [PubMed]

horticulturae

MDPI

Article

Efficient Micropropagation Using Different Types of Explant and Addressing the Hyperhydricity of *Ballota acetabulosa*, a Mediterranean Plant with High Xeriscaping Potential

Georgia Vlachou and Maria Papafotiou *

Laboratory of Floriculture and Landscape Architecture, Department of Crop Science, School of Plant Sciences, Agricultural University of Athens, Iera Odos 75, 11855 Athens, Greece; g_vlaxou@hotmail.com
* Correspondence: mpapaf@aua.gr; Tel.: +30-210-5294-555

Abstract: *Ballota acetabulosa* (L.) Benth. (syn. *Pseudodictamnus acetabulosus* (L.) Salmaki and Siadati), f. Lamiaceae, the Greek horehound, is a compact evergreen small shrub native to Greece, with hairy grey-green leaves, that bears small pink-purple flowers with green conical calyxes along its erect stems in late spring. The species stands out for its high resistance in xerothermic conditions and therefore it is advisable to promote its use in xeriscaping. The aim of this study was to develop an efficient protocol for in vitro propagation of *B. acetabulosa* for introduction into the horticultural and pharmaceutical industries. Shoot tip and single node explants derived from in vitro seedlings were cultured on MS medium with various cytokinin types and concentrations. Explants responded at almost 100% to produce high number of shoots on a medium with 1.0 mg L^{-1} zeatin or 6-benzyladenine. However, there was intense hyperhydricity in the cultures, which was addressed in further experiments by increasing agar concentration from 8 to 12 g L^{-1}, preserving high multiplication indices (92% response, 10.2 shoots per explant). Microcuttings with 2–3 visible nodes, either from the apical part, including the apical meristem, or from the basal part of microshoots, as well as microshoot clusters, rooted 100% on full- or half-strength MS medium, respectively, regardless of the addition of indole-3-butyric acid (IBA, 0.5–4.0 mg L^{-1}) in the rooting medium. However, middle level concentrations of IBA increased the number and length of roots produced, while the higher its concentration, the more and longer axillary shoots developed in the microcuttings during the rooting period. The acclimatization of all plantlets was completely successful (100%) in ex vitro conditions on peat/perlite substrate (1:1, v/v). Thus, efficient methods of producing propagation material to promote *Ballota acetabulosa* as a horticultural and medicinal plant were developed. In particular, rooting of microshoot clusters or microcuttings without the shoot tip, in the presence of 1.0 mg L^{-1} IBA, leads to a plant of suitable shape for the floricultural market, without the need for further manipulation (pruning) in the nursery.

Keywords: agar concentration; cytokinin type; Greek horehound; in vitro propagation; in vitro rooting by nodal cuttings; microshoot cluster rooting; proliferation in successive subcultures; *Pseudodictamnus acetabulosus*; xerophyte

1. Introduction

Ballota acetabulosa (L.) Benth., syn. *Pseudodictamnus acetabulosus* (L.) Salmaki and Siadati (f. Lamiaceae), the Greek horehound, is a compact, bushy, evergreen xerophyte, 30–60 cm height, being one of the most widely spread small shrubs all over Greece, especially in the SE, including Crete and other Greek islands, except the area of the Pindos mountain range in the northwest [1]. It is also native in western Turkey [2]. It is found mainly in poor alkaline soils, dry lands, and stony places [3–5], but grows equally well in sandy and clay soils. The plant belongs to the group of late-spring/summer-flowering species of Greece [6]; has erect-white-felted stems and paired, ovate, woolly grey-green heart-shaped leaves; and bears small pink-purple flowers with green conical calyxes along the stems in the axils of the leaves. The characteristic epithet of the genus "acetabulosa" comes from the Latin acetabulum = calyx, since the appearance of the plant is characterized by the presence of many dense calyces (Figure 1).

Figure 1. *Ballota acetabulosa* young plant in xeriscaping in late March (**A**), shoot sprouting and leaf (**B**), upper (**C**) and lower (**D**) leaf surface, plant growing wild in Mount Hymettus, Athens (Attika) in May (**E**), shoots with the dense calyces of the plant in July (**F**), stem at flowering in late August (**G**), plant in November where dried shoot inflorescence and new vegetation are showing (**H**).

Since ancient times, people have used *B. acetabulosa* for many purposes [7]. The Greek Pedanius Dioscorides (c. 40–90 AD), "the father of pharmacognosy", describes the plant and its medicinal properties in *De materia medica* (On Medical Material), a Greek encyclopedia on herbal medicine and related medicinal substances, which was widely read for more than 1500 years. Nowadays, the species retains interest as a medicinal plant, as it has mild astringent and antibacterial properties and is still used for nausea, vomiting, dyspepsia, and as a sedative [8–10]. Also, methanol extracts of *B. acetabulosa* have been reported to possess antioxidant activity [11]. The seed vessels were used by Greek peasants as floating wicks in lamps of olive oil mainly in front of icons, and the plant is therefore sometimes called "louminia".

The appearance of the plant and its tolerance to barren, dry, alkaline soils and xerothermic conditions make it suitable to be introduced in xeriscaping (Figure 1A) and sustainable green spaces in semi-arid or arid areas with Mediterranean climate. The lack of aggressiveness makes it suitable for use in archaeological sites as well [12]. As it attracts pollinators and has limited needs for irrigation serves the current trend for sustainable landscaping. The conservation and promotion of pollinator biodiversity is promoted in the urban landscape, through the planting of flowering plants, especially native perennials [13].

There are no reports found in the literature on the propagation of *B. acetabulosa*, apart from a preliminary publication of ours on the micropropagation of the species [14]. The work revealed hyperhydricity problems that prevented the proliferation from reaching a satisfactory level for commercial use. Hyperhydricity is a morphological and physiological disorder that occurs in in vitro cultures [15], resulting in abnormal plant growth, reduced proliferation potential, poor plant quality, failure to root, and reduced ability to acclimatize ex vitro [16,17]. Factors shown to be responsible for hyperhydricity are the gelling agent, cytokinins [18], ammonium ions [19], and culture ventilation [20].

Seedlings have been shown to be a parental material that ensures successful in vitro culture establishment and high proliferation indices, especially in native Mediterranean xerophytic species [21–25]. Furthermore, when native plants are to be used in natural landscape restorations, their propagation by seed is indicated, as this enhances the genetic diversity desired for the natural landscape. In addition, the genetic variability that can exist using seedlings as a parent material for propagation contributes to the selection of genotypes with high medicinal value [26].

The aim of this study was to develop an efficient protocol for in vitro propagation of *B. acetabulosa*, starting from seedlings derived from in vitro germinated seeds, for the wider utilization of the species (a) as an ornamental plant, particularly for xeriscaping, landscape restoration, sustainable urban green spaces, and archaeological parks and (b) as a medicinal plant. Specifically, we studied the (a) effect of type and concentration of cytokines and agar concentration on the in vitro proliferation and elimination of hyperhydric problems, (b) effect of explant type on proliferation, (c) effect of auxin and microcutting type on in vitro rooting, and (d) ex vitro acclimatization of rooted microcuttings and microshoot clusters. Our ultimate goal is to have an efficient method for producing propagation material to promote the species as a horticultural and medicinal plant.

2. Materials and Methods

2.1. Effect of Cytokinin Type and Concentration on In Vitro Culture Establishment

In vitro cultures were established from in vitro germinated seedlings. Seeds were harvested at the stage of full maturity in August from a single native *B. acetabulosa* adult plant located on the western side of Ancient Corinth (Corinth, 37°53′58.5″ N and 22°52′38.5″ S). Two months after harvest, seeds were placed for germination in vitro, after being surface-sterilized by an aqueous solution of 15% v/v commercial bleach (4.5% w/v sodium hypochloride) with 1–2 drops of Tween 20 (polyxyethylenesorbitan monolaurate, MERCK KGaA, Darmstadt, Germany) for 15 min. Seeds were sown in plastic Petri dishes on hormone-free (Hf), half-strength MS medium [27] containing 20 g L^{-1} sucrose and solidified with 8 g L^{-1} agar (ROUMBOULAKIS S.A. Athens, Greece), at 15 °C and a 16 h photoperiod of cool white fluorescent light (37.5 μmol m^{-2} s^{-1}). The medium pH was adjusted to 5.7 before agar addition and autoclaving (121 °C, 20 min). After the completion of germination (at 20 days), the seedlings were transferred to full-strength MS medium for further growth for 40 days to be used as explant source for the initial in vitro culture.

The initial culture was established by single-node explants, excised from two-month-old in vitro-grown seedlings (two to four explants from each seedling). Explants were cultured on solid (8 g L^{-1} agar) MS medium with 30 g L^{-1} sucrose, either hormone-free (Hf-MS) or supplemented with either zeatin (ZEA), 6-benzyladenine (BA), kinetin (KIN), or 6-(γ,γ-Dimethylallylamino)Purine (2iP) at a concentration of 0.5, 1.0, or 2.0 mg L^{-1}. All chemicals used in this research were from MERCK KGaA, Darmstadt, Germany.

2.2. Effect of Explant and Cytokinin Type and Concentration on Shoot Multiplication

Shoot-tip or single-node explants (0.6 cm long) were excised from microshoots that had been produced at the establishment stage, 42 d after initial planting, and put on the same medium as they originated from (i.e., Hf-MS, MS medium supplemented with 0.5, 1.0, or 2.0 mg L^{-1} ZEA, BA, KIN, or 2iP). The effect of cytokinin type and concentration and that of explant type (shoot tip or nodal) were determined. In subsequent subcultures (2nd, 3rd, 4th), shoot-tip or single-node explants (0.6 cm long) were excised from microshoots and cultured on either Hf-MS or MS medium supplemented with 1.0 mg L^{-1} BA or ZEA. The explants were placed each time on the same medium as they originated from.

2.3. Effect of Agar Concentration on Hyperhydricity

In order to address the hyperhydricity problem that appeared at the shoot multiplication stage, increasing the concentration of agar was tested. Therefore, in the 5th subculture, shoot-tip or single-node explants excised from microshoots grown on MS medium with 1 mg L^{-1} BA were cultured either on Hf-MS solidified with 8 g L^{-1} agar or MS medium supplemented with 0.5 or 1.0 mg L^{-1} BA and solidified with 8 or 12 g L^{-1} agar, in all possible combinations.

2.4. In Vitro Rooting

For in vitro rooting, microcuttings (2.0–2.5 cm long) either from the apical part including the apical meristem or from the basal part of microshoots were used (Figure 2A). The microshoots from which the microcuttings were derived were about 5 cm long. Microcuttings were placed for rooting on either Hf-MS or MS medium with 0.5, 1.0, 2.0, or 4.0 mg L^{-1} indole-3-butyric acid (IBA) for 6 weeks. Also, microshoot clusters (Figure 2B) were placed for rooting on either half-strength Hf-MS medium or half-strength MS with 0.5, 1.0, 2.0, or 4.0 mg L^{-1} IBA for 6 weeks.

Figure 2. *Ballota acetabulosa* microcuttings from the apical (**A**) and the basal (**B**) part of the microshoot and clusters of microshoots (**C**) used for in vitro rooting.

2.5. In Vitro Culture Conditions and Data Collection

In vitro cultures were carried out in 100 mL Sigma glass vessels covered by Magenta B-Caps, with 25 mL medium (four explants per vessel in establishment and shoot multiplication stages, two shoot clusters or four shoot tip or nodal stem cuttings per vessel in rooting experiments). The incubation of all in vitro cultures was performed at 25 °C under a 16 h photoperiod of 37.5 μmol m^{-2} s^{-1} photosynthetic photon flux density of cool-white fluorescent light. The medium pH was adjusted to 5.7 before agar and autoclaving (121 °C, 20 min). Data were recorded after 42 d of culture.

In shoot multiplication experiments, data were recorded on (a) the percentage of explants that responded for shoot production [total shooting (%)], which was divided into three parts: (i) shooting[1] (%): explants formed both normal shoots (NSh) and hyperhydric shoots (HSh); (ii) shooting[2] (%): explants formed only hyperhydric shoots; and

(iii) shooting[3] (%): explants formed only hyperhydric mass (shoot and node numbers could not be recorded); (b) NSh and HSh number per explant; (c) shoot length of NSh, for explants that simultaneously produced NSh and HSh; and (d) multiplication index: shooting[1] × NSh number × NSh node number. In rooting experiments, rooting percentage and root number and length were recorded.

Twenty-four explants per treatment (6 replicates) were used in the establishment and multiplication stages, while in the rooting experiments, sixty microcuttings (15 replicates) and forty shoot clusters (20 replicates) were used per treatment.

2.6. *Ex Vitro Acclimatization and Establishment*

Rooted microshoots and shoot clusters after being rinsed thoroughly with running tap water to remove growth medium were transferred to 1: 1 (v/v) peat/perlite substrate (peat: high-moor with adjusted pH of up to 5.5–6.5, Klasmann-Delimann Gmbh, Geeste, Germany; perlite: particle diameter 1–5 mm, Perloflor, ISOCON S.A., Athens, Greece). Plastic containers 500 mL were used with eight plantlets per container, covered with a transparent plastic film (SANITAS, Sarantis S.A., Athens, Greece), and placed in a growth chamber for 7 d, at 20 °C with a 16 h photoperiod at 37.5 μmol m^{-2} s^{-1}, provided by cool-white fluorescent lamps. After 7 d, the containers were transferred in a heated glasshouse (37°58′58.0″ N, 23°42′19.2″ E), in the mist (substrate temperature 22 °C maintained by thermostatically controlled electric heating cable), where they remained for 10 d and then were placed on a greenhouse bench for another 30 d, at the end of which data for acclimatization were recorded. The total number of plantlets transferred ex vitro from both rooted microcuttings (apical and basal) and rooted microshoot clusters was 120 (15 containers, 8 plantlets per container).

After acclimatization was completed, plantlets were transplanted singly into plastic pots (500 mL) with peat/perlite 2:1 v/v and fertilized monthly with 2.0 g L^{-1} complete water-soluble fertilizer (Nutrileaf 60, 20-20-20; Miller Chemical and Fertilizer Corp., Hanover, PA, USA). Plant establishment data were recorded four months later.

2.7. *Statistical Analysis*

A completely randomized design was used in all experiments. The significance of the results was tested by one-, two-, or three-way analysis of variance (ANOVA) (F test, discrete variables followed the normal distribution, arcsine transformation was used for the data of explant response for shoot formation and rooting percentage prior to statistical analysis) and the means of the treatments were compared using Student's t test at $p \leq 0.05$ (JMP 13.0 software, SAS Institute, Cary, NC, 2013, USA).

3. Results

3.1. *In Vitro Culture Establishment*

At the culture establishment stage of single-node explants from in vitro seedlings, a significant interaction of the experimental factors was recorded in most of the parameters measured, except the number of normal shoots and the node number of normal shoots, which were affected by both the type and the concentration of the cytokinin used (two-way ANOVA, Table S1 in Supplementary Materials). Explants were successfully established in vitro even on Hf-MS, in which they developed only one or two axillary shoots that had however the greatest elongation compared to those induced in cytokinin-containing media (Figure 3A–C). In all media, more than 80% of the explants produced shoots, while the response reached 100% in those with 0.5 mg L^{-1} KIN, 1 mg L^{-1} ZEA or BA, or 2 mg L^{-1} ZEA or KIN. The highest number of shoots per explant was recorded in all ZEA and BA concentrations and in 2.0 mg L^{-1} KIN, which were much shorter and with less internodes compared to those in the Hf-MS (Figure 3A–D and Figure 4). However, increasing the cytokinin concentration in the medium from 0.5 to 1 or 2 mg L^{-1} caused an increase in hyperhydricity (Figure 3A,B), but the media with 1 mg L^{-1} ZEA or 2.0 mg L^{-1} KIN still gave the highest multiplication index of normal shoots (Figure 3E).

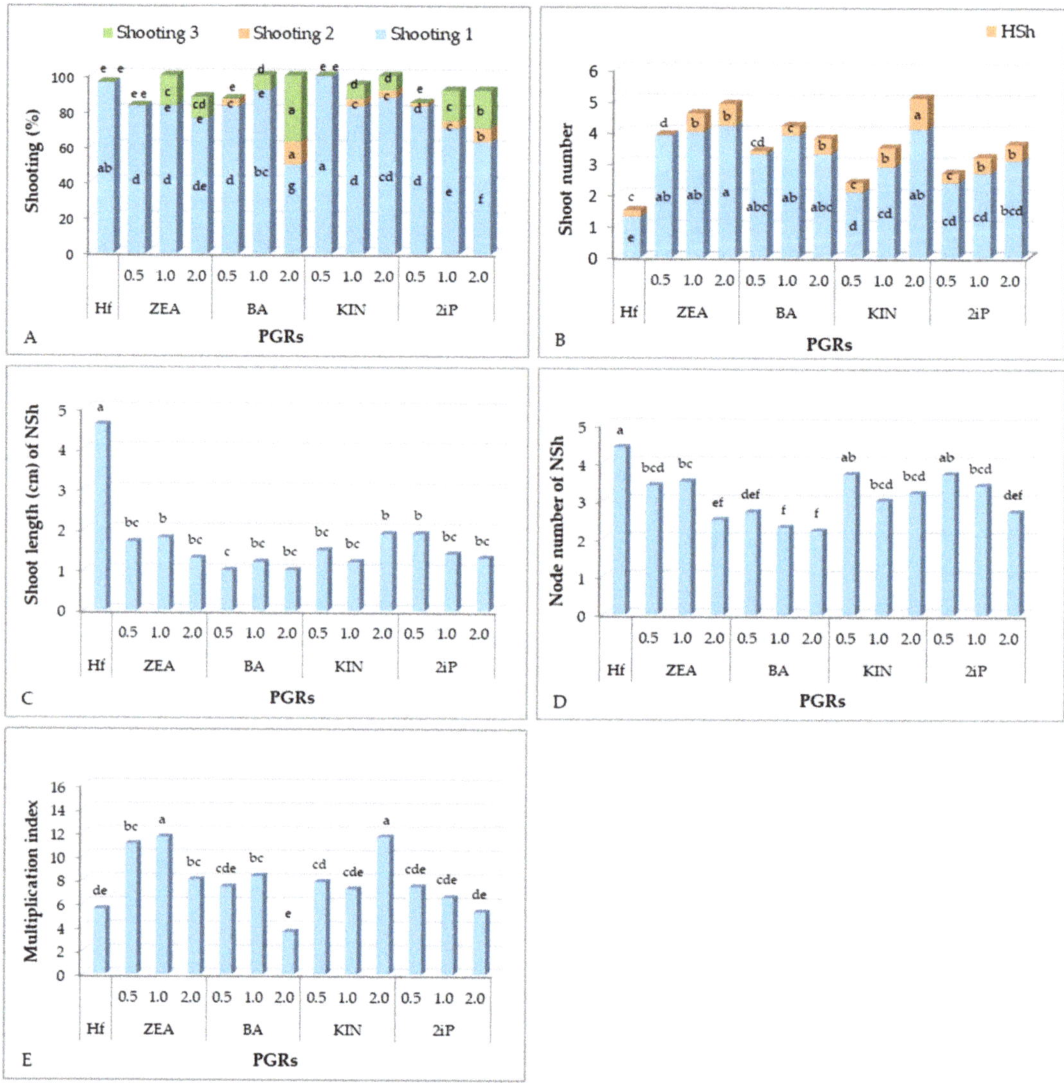

Figure 3. Effect of cytokinin type and concentration on the response of *Ballota acetabulosa* single-node explants, derived from in vitro grown seedlings, at the culture establishment stage. One-way ANOVA, mean separation by Student's t at $p \leq 0.05$, $n = 24$. Means followed by the same letter a–f do not defer at $p \leq 0.05$. Shooting[1]: the explants gave normal (NSh) and hyperhydric shoots (HSh). Shooting[2]: the explants gave hyperhydric shoots only. Shooting[3]: the explants gave a hyperhydric mass. Multiplication index = shooting[1] (%) × shoot number of NSh × node number of NSh. The blue color in figures refers to data for normal shoots.

Figure 4. Response of *Ballota acetabulosa* single-node explants, derived from in vitro grown seedlings, at the culture establishment stage, on Hf-MS medium (**A**), or MS medium supplemented with 1.0 mg L^{-1} ZEA (**B**), 2.0 mg L^{-1} BA (**C**), 2.0 mg L^{-1} KIN (**D**), or 2.00 mg L^{-1} 2iP (**E**). Bar size: 1 cm.

3.2. Shoot Multiplication at First Subculture

In the first subculture, where both shoot-tip and single-node explants were excised from microshoots produced at the establishment stage and cultured in the same media from which the explants were derived, there were significant interactions of experimental factors, except in the case of the normal shoot number and length, which were not affected by explant type and cytokinin concentration, respectively (three-way ANOVA, Table S2 in Supplementary Materials). Thus, we proceeded to two-way ANOVA for shoot-tip and single-node explants separately, which showed significant effects of cytokinin type on shoot length and node number in both explant types (Table S3 in Supplementary Materials) and a significant effect of cytokinin concentration on node number only in single-node explants (Table S4 in Supplementary Materials).

The explant response for shoot production was similar to that of the establishment stage; however, hyperhydricity was increased and in most media was more pronounced in single-node explants (Figure 5A,B). Shoot number was highest in media with 1 mg L^{-1} ZEA or BA and the highest of all was observed in single-node explants in the medium with 1 mg L^{-1} ZEA (eight normal shoots plus two hyperhydric per explant, Figures 5C,D and 6). However, shoot length and number of nodes were reduced in the presence of cytokinin, particularly in ZEA and BA (Figures 5E,F and 6), where the percentage of explants producing normal shoots was also reduced due to increased hyperhydricity (Figure 5A,B), and this resulted in multiplication indices similar to those at the establishment stage, although shoot numbers per explant were higher at the subculture (Figures 3B,E and 5C,D,G).

3.3. Effect of Continuous Subcultures on Shoot Multiplication

In four consecutive subcultures of shoot-tip or single-node explants on media either Hf or supplemented with 1 mg L^{-1} ZEA or BA, factorial analysis showed significant interactions between the three experimental factors, i.e., subculture number, cytokinin type, and explant type, and thus two-way ANOVAs were determined separately for shoot-tip or single-node explants, which showed an effect of PGRs on shoot length that was not affected by the subculture number when shoot-tip explants were used (Table S5 in Supplementary Materials) and an effect of both PGRs and subculture number on multiplication index in case of single-node explants (Table S6 in Supplementary Materials).

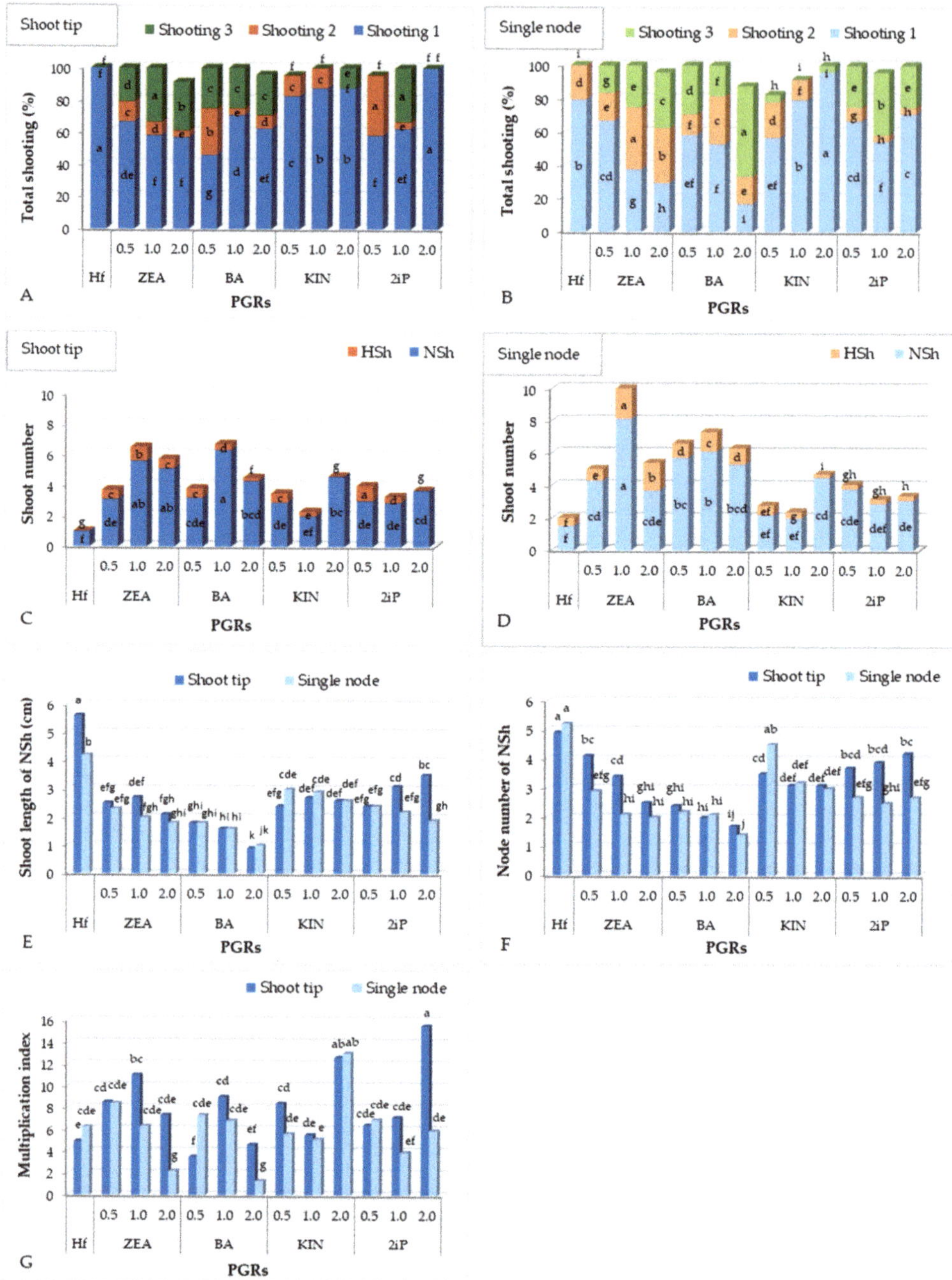

Figure 5. Effect of explant type, cytokinin type, and cytokinin concentration on the response of *Ballota acetabulosa* shoot-tip or single-node explants at the first subculture. One-way ANOVA, mean separation by Student's t at $p \leq 0.05$, $n = 24$. Means followed by the same letter a–k do not defer at $p \leq 0.05$. Shooting^{1}: the explants gave normal (NSh) and hyperhydric shoots (HSh). Shooting^{2}: the explants gave hyperhydric shoots only. Shooting^{3}: the explants gave a hyperhydric mass. Multiplication index = shooting^{1} (%) × shoot number of NSh × node number of NSh.

Figure 6. Response of *Ballota acetabulosa* shoot tip or single-node explants at the 1st subculture on Hf-MS medium (**A**), or MS medium supplemented with 1.0 mg L^{-1} ZEA (**B**), or BA (**C**). Bar size: 1 cm.

When shoot-tip explants were used, the percentage of explants that responded to shoot production remained constant in subcultures on Hf or ZEA medium, whereas it decreased on BA medium as subcultures progressed. In the case of single-node explants, as subcultures progressed, there was a decrease in explant response in all media (Figure 7A,B). Regarding hyperhydricity, in most cases, it increased as the subcultures progressed (Figure 7A,B). More shoots were produced in cytokinin media, and in these media, there was an increase in the number of shoots in the third and fourth subculture, while shoot length and number of nodes remained constant in the subcultures, being lower than in Hf medium (Figures 7C–F and 8). The large increase in shoot number in the third and fourth subcultures in cytokinin media did not lead to an analogous increase in multiplication indices due to the simultaneous increase in the number of hyperhydric shoots per explant (Figures 7C,D,G and 8). Thus, addressing hyperhydricity proved necessary.

3.4. Effect of Agar Concentration on Hyperhydricity

A fifth subculture took place using shoot-tip and single-node explants mixed where MS medium was supplemented with BA at 0.5 or 1.0 mg L^{-1} and solidified either as usual with 0.8 g L^{-1} agar or with 12 g L^{-1} agar, while Hf-MS solidified with 0.8 g L^{-1} agar was also used as a control. Greatest number of shoots was produced with the 1 mg L^{-1} BA medium. Hyperhydricity, which was higher in BA media, was eliminated to the control level when the agar concentration was increased. Thus, media with BA and solidified with 12 mg L^{-1} agar gave high multiplication indices (Table 1 and Figure 9).

3.5. In Vitro Rooting

Apical and basal microcuttings showed 100% rooting on MS medium in all treatments even those without IBA. The addition of IBA up to 1 mg L^{-1} or 2.0 mg L^{-1} to the medium increased the number and length of roots in the apical or basal microcuttings, respectively, while at 4 mg L^{-1}, the response was as in the control (Figure 10A,B). Concurrent with rooting, axillary bud sprouting was observed in IBA media, which was more pronounced in basal microcuttings and high IBA concentrations (Figure 11).

Microshoot clusters, like the microcuttings, rooted 100% in all treatments, showing an increase in the number of roots produced per cluster proportional to the increase in IBA concentration, up to the highest IBA concentration tested, i.e., 4 mg L^{-1} (Figure 12A). Regarding root length, it was highest at 0.5 mg L^{-1} IBA and lowest at 2.0 and 4.0 mg L^{-1} IBA (Figure 12B). At the same time, elongation of the microshoots was observed in all media, with the greatest elongation occurring in some of the microshoots in the Hf medium (Figure 13).

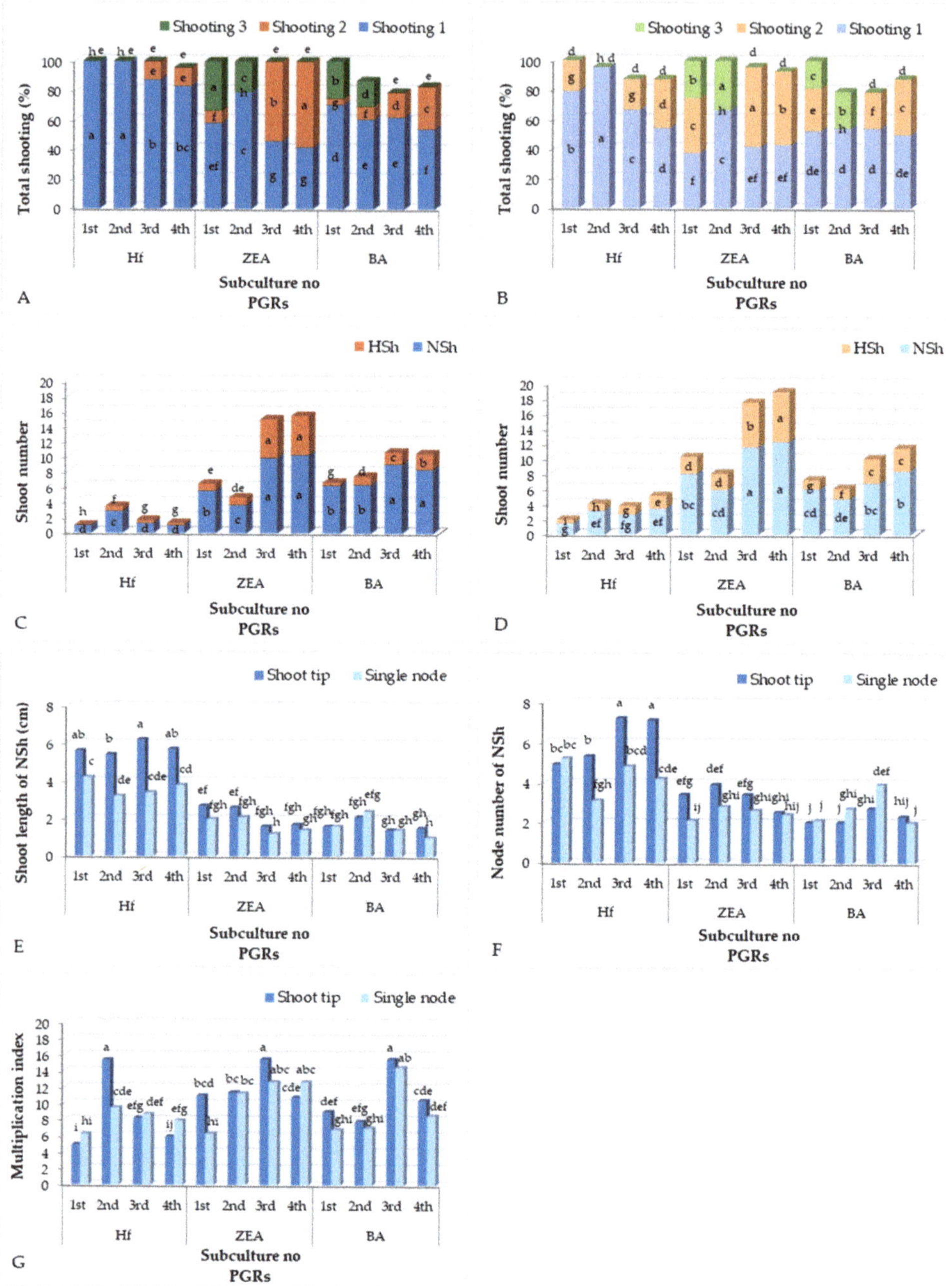

Figure 7. Effect of BA and ZEA at 1 mg L^{-1} on the response of *Ballota acetabulosa* shoot-tip or single-node explants in continuous subcultures at the multiplication stage. One-way ANOVA, mean separation by Student's t at $p \leq 0.05$, $n = 24$. Means followed by the same letter a–j do not defer at $p \leq 0.05$. Shooting^1: the explants gave normal (NSh) and hyperhydric shoots (HSh). Shooting^2: The explants gave hyperhydric shoots only. Shooting^3: The explants gave a hyperhydric mass. Multiplication index = shooting^1 (%) × shoot number of NSh × node number of NSh.

Figure 8. Response of *Ballota acetabulosa* shoot tip or single-node explants at the 3rd subculture on Hf-MS medium (**A**), or MS medium supplemented with 1.0 mg L^{-1} ZEA (**B**) or BA (**C**). Bar size: 1 cm.

Table 1. Effect of BA and agar concentration on the response of *Ballota acetabulosa* shoot-tip and single-node explants (mixed) in the 5th subculture.

BA/Agar (mg L^{-1})/ (g L^{-1})	Shooting (%) [1/2] (%)	Number of NSh [T]/HSh [TT]	Length of NSh (cm)	Node Number of NSh	Multiplication Index
0.0/8 [±]	88 a [z]/8 c	1.6 d/0.1 e	5.4 a	5.9 a	8.3 b
0.5/8	68 b/27 a	4.7 c/2.3 b	1.4 b	2.4 b	7.7 b
0.5/12	88 a/8 c	7.8 b/1.0 cd	1.3 b	2.3 b	15.8 a
1.0/8	70 b/22 b	7.4 b/3.4 a	1.1 bc	1.9 bc	9.8 b
1.0/12	92 a/8 c	10.2 a/0.8 d	0.8 c	1.8 c	16.9 a
F_{BA}	NS/-	***/***	***	**	NS
F_{agar}	***/-	***/**	**	NS	***
$F_{BA \times agar}$	NS/***	NS/NS	NS	NS	NS
$F_{one\text{-}way\ ANOVA}$	***/***	***/***	***	***	***

[±] Data excluded from 2-way ANOVA; [z] mean separation in columns by Student's t, $p \leq 0.05$. NS or **, ***, non-significant at $p \leq 0.05$ or significant at $p \leq 0.01$ or $p \leq 0.001$, respectively; $n = 24$. Means followed by the same letter a–e do not defer at $p \leq 0.05$. Multiplication index = shooting[1] (%) × shoot number of NSh × node number of NSh. [1] The explants produced normal and hyperhydric shoots. [2] The explants produced hyperhydric shoots only. [T] NSh = normal shoot. [TT] HSh = hyperhydric shoot.

Figure 9. Response of *Ballota acetabulosa* shoot-tip and single-node explants (mixed) in the 5th subculture on MS medium with BA and agar concentrations (mg L^{-1} and g L^{-1}, respectively) shown. Bar size: 1 cm.

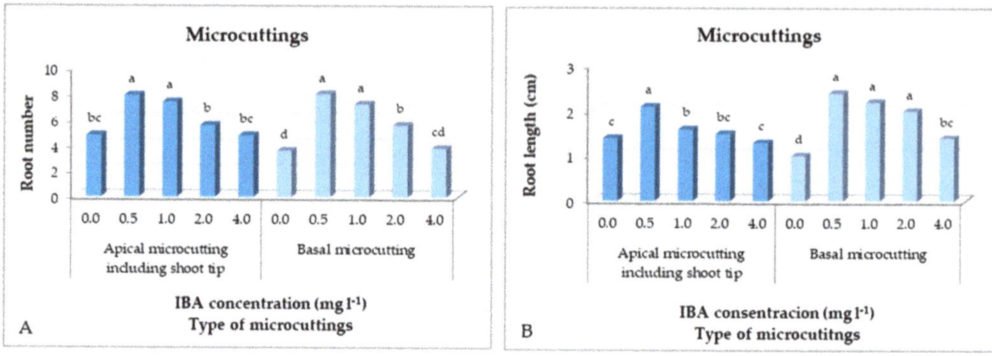

F_{IBA}	***	**
$F_{microcutting}$	NS	NS
$F_{IBA \times microcuttings}$	NS	NS
$F_{one\text{-}way\ ANOVA}$	***	***

Figure 10. Effect of IBA concentration and microcutting type on root number (**A**) and root length (**B**) during in vitro rooting of *B. acetabulosa* on MS medium (rooting response 100% in all treatments). Mean separation in each figure by Student's t, $p \leq 0.05$. Means followed by the same letter a–d do not defer at $p \leq 0.05$. NS or **, ***: non-significant at $p \leq 0.05$ or significant at $p \leq 0.01$, $p \leq 0.001$, respectively; n = 60.

Figure 11. In vitro rooting of apical or basal microcuttings of *B. acetabulosa* on MS medium under the effect of IBA concentrations shown. Bar size: 1.0 cm.

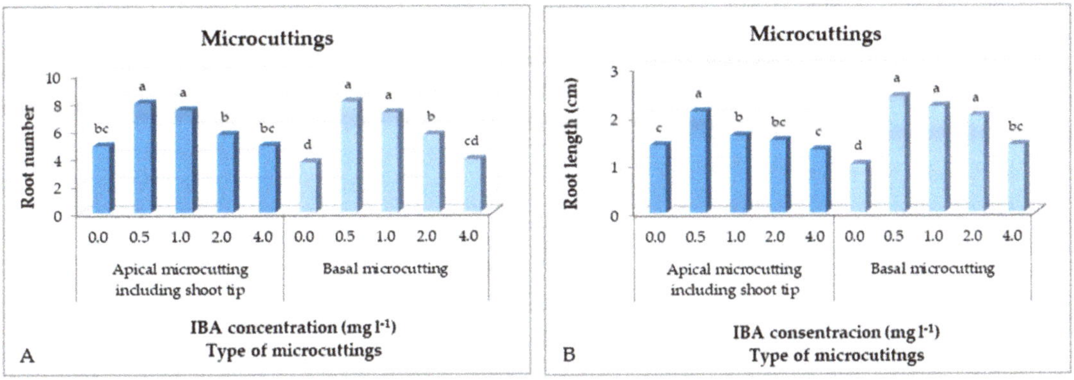

Figure 12. Effect of IBA concentration on root number (**A**) and root length (**B**) during in vitro rooting of *B. acetabulosa* microshoot clusters on half-strength MS medium (rooting response 100% in all treatments). Mean separation in each figure by Student's t, $p \leq 0.05$. Means followed by the same letter a–d do not defer at $p \leq 0.05$. $F_{\text{one-way ANOVA}}$ significant at $p \leq 0.001$, $n = 40$.

Figure 13. In vitro rooting of *B. acetabulosa* microshoot clusters on half-strength MS medium under the effect of IBA concentration shown. Bar size = 1.0 cm.

3.6. Ex Vitro Acclimatization

Rooted microcuttings of both types, apical and basal, and rooted clusters of microshoots were acclimatized ex vitro at 100%, maintaining the characteristics of the mother plants. Further plant growth for the next three months after transplanting was 100% successful (Figure 14A,B).

Figure 14. Rooted *B. acetabulosa* microshoot cluster transplanted on peat/perlite (1:1, *v*/*v*) for acclimatization (**A**) and rooted microshoot clusters three months after their ex vitro transplantation (**B**).

4. Discussion

4.1. In Vitro Culture Establishment

The floriculture and landscape plant industry is constantly looking for new products to introduce to markets. In addition, the climate crisis is a new parameter that forces the investigation of new species with tolerance to harsh environments so that they can withstand and show satisfactory growth in xerothermic conditions, water scarcity, sudden changes in environmental conditions, or increased urban pollution [28,29]. In arid and semi-arid regions, research on indigenous species with tolerance to heat and drought and the ability to grow with minimal water supply is very important in order to introduce them into horticulture and landscape architecture, as water is a precious commodity and must to be provided sparingly in urban greenery [30–33].

B. acetabulosa is a promising for xeriscaping native plant of Greece and western Turkey [1,2], and an efficient micropropagation protocol will be useful in its commercial production process. Further, given its medicinal properties [10,13], in vitro propagation will facilitate the isolation of clones with desirable characteristics.

The establishment of in vitro cultures from seedlings developed in vitro ensures sterile mother material, overcoming the obstacle of surface sterilization of the explants, which in the case of plants with woolly tissues such as *B. acetabulosa* is rather difficult [34,35]. Moreover, explants from young in vitro seedlings often respond more readily to in vitro culture compared to similar explants from adult plants [24,36]. In addition, this can lead to the production of clones with special characteristics, while the variability of cultures ensures a level of diversity that is desirable in the use of plants in landscape restoration. To the best of our knowledge, there is no documentation in the literature on the micropropagation of *B. acetabulosa*, apart from a preliminary publication by us [14].

In the present work, the initial establishment of cultures from in vitro grown seedlings was easy and successful. *Ballota acetabulosa* is an open-pollinated species; thus, even seeds from the same plant could have genetic variability. However, no variability was recorded in terms of in vitro seedling growth and the response of seedling explants in the following experiments, e.g., in the Hf-control medium at the culture establishment stage, the seedling explants showed 98% response and formed 1–2 shoots each. Nodal explants reacted at rates higher than 80% even in Hf-MS medium, where they gave 1–2 shoots per explant, while the addition of all four cytokinin types tested induced a sufficient number of shoots per explant with satisfactory elongation, distinguishing ZEA and BA at a concentration of 1 mg L^{-1}, where approximately 4 shoots were formed per explant. Similarly, highly successful establishment of in vitro cultures from aseptically germinated seedlings has been shown in other xerophytic species of the family Lamiaceae, such as the Mediterranean *Salvia* spp.: *S. fruticosa*, *S. officinalis*, *S. ringens*, and *S. pomifera* ssp. *pomifera* [25], *Teuctium capitatum* [22] and *Sideritis athoa* [37], as well as Mediterranean xerophytes of other families, e.g., *Dianthus fruticosus* [38] and *D. cruentus* of f. Caryophyllaceae [23,39], *Lithodora zahnii* f. Boraginaceae [40], and *Anthyllis hermanniae* and *Ebenus sibthorpii* f. Fabaceae [24,41]; however, there are exceptions, such as, e.g., Lamiaceae *S. tomentosa*, which did not show high establishment rates from aseptically germinated seedlings compared to greenhouse-grown adult plants [35]. Micropropagation of native species using explants from sterile juvenile tissues, in addition to ensuring a lack of microbial contaminants, increases response rates and the absence of recalcitrance, offering genetic variability that is desirable when the plants will be used for landscape restoration, as in nature, the introduction of a single clone is not desirable [42]. Thus, in vitro culture of pre-existing apical or nodal meristems is the most reliable and widely used method for micropropagation of many species, including medicinal and aromatic plants [43].

4.2. Multiplication Stage and Addressing Hyperhydricity

In the first subculture, which was carried out on substrates identical to those of the initial culture, in addition to single-node explants, shoot-tip explants were also used. The response percentage of the explants to produce normal shoots was reduced as the hyperhydricity percentage of the explants intensified compared to the initial culture; however, the number of shoots per explant increased, and at 1 mg L^{-1} ZEA or BA, it reached 5.5 or 6 in the shoot-tip explants and 7.8 or 6.8 in the single-node explants, respectively, while the microshoots produced showed greater elongation compared to the initial culture. In the third and fourth subculture in 1 mg L^{-1} ZEA or BA, the number of normal shoots increased to 9–11 per explant in ZEA and 6–8 in BA, while the hyperhydric shoots also increased up to 4 or 6 per explant in BA and ZEA, respectively. The increased number of shoots as well as the hyperhydricity in successive subcultures on the same medium may be due to a cytokinin accumulation effect in plant tissues. In the four consecutive subcultures, both types of explant responded in a rather similar manner, except that hyperhydricity in most cases was more pronounced in single-node explants compared to shoot tip ones, verifying previous results concerning Mediterranean *Salvia* spp. [25,35]. Increased numbers of shoots in a subculture carried out on the same media as those of the establishment stage were also shown in the Mediterranean xerophytes *A. hermaniae* [24], *Salvia* spp. [25], and *L. monopetalum* [44], while in others such as *D. cruentus* [23], *E. sibthorpii* [41], and *Clinopodium nepeta* [36], shoot proliferation in the subculture remained rather constant.

In tissue cultures of herbaceous plants and especially in cultures derived from seedlings, the problem of hyperhydricity is often encountered, mainly at the stage where increased concentrations of cytokinins are used to induce high shoot proliferation, thus negatively affecting growth quality and causing significant loss of plant material. Hyperhydricity has been reported in micropropagation using seedlings as parent material in several Mediterranean xerophytes, e.g., *Globularia alypum* [45], *Lithodora zahnii* [40], *Sideritis athoa* [37], *Anthyllis barbajovis* [46], *Salvia tomentosa* [35], and *Dianthus cruentus* [23]. As mentioned above, hyperhydricity in general was more pronounced in single-node explants compared to shoot tip ones, verifying previous results with other Mediterranean xerophytic species [25,35]. The different endogenous hormonal statuses of the shoot-tip and single-node explants combined with the exogenous effect of the substrate cytokinin probably caused different intensities of hyperhydration. Alternatively, or simultaneously, the presence of the shoot tip in the explants may have affected the absorption of substrate components and water in a different way than in the nodal explants. Hyperhydricity can be due to various factors, such as genotypic sensitivity, reduced aeration of cultures and ethylene accumulation in the culture vessels [20,47,48], nitrate and ammonium composition in the culture medium [19,47,49], type and amount of cytokinin [18,19,48,50,51], and cytokinin/auxin balance in the medium [36,52,53]. Taking into account these factors, which contribute to the complexity of the phenomenon of hyperhydricity, addressing hyperhydricity must be species-specific [51].

In the present study, increasing agar concentration eliminated hyperhydricity without negatively affecting shoot proliferation indices. On the contrary, the multiplication index increased. These findings verified similar results regarding the treatment of hyperhydricity by increasing agar concentration in the micropropagation of other Mediterranean species, e.g., the xerophytes *Lithodora zahnii* [40], *Globularia alypum* [45], *C. nepeta* [36], *Calamintha cretica* [52], and *Dianthus cruentus* [23], as well as other herbaceous species [54,55]. Modifying the type and concentration of the solidifying agent in the medium regulates the water potential, the diffusion of micronutrients, and the uptake of cytokinins by the explants [56].

4.3. In Vitro Rooting

Rooting of in vitro-produced microshoots is a very important stage in micropropagation. The goal is rapid rhizogenesis and elongation of the roots in order to achieve a strong and rich root system that will subsequently ensure successful acclimatization to ex vitro conditions. The most common procedure is to place the tops of microshoots 1–3 cm long in half-strength MS medium containing some amount of auxin, usually IBA. In our work, rooting of microcuttings in MS medium was completely successful, even without the use of auxin; however, the addition of auxin induced more and longer roots. The use of seedlings as mother plants for in vitro cultures establishment probably facilitated in vitro response. The ease of rooting confirms the generally accepted fact that juvenile tissues are most responsive in culture compared to mature tissues [57,58]. It is noteworthy that microcuttings from the basal part of the microshoots rooted equally successfully to apical microcuttings that included the tip of the shoot, the latter being the standard procedure for in vitro rooting. The basal microcuttings produced roots with shoots simultaneously. This is a positive finding because the entire length of a long microshoot can be used for producing more than one plantlet. Placing single-node pieces of a microshoot for rooting results in more rooted plantlets per microshoot, and in this way, costs in commercial production could be reduced. To our knowledge, in vitro rooting of stem pieces, without apical meristem, with simultaneous shoot production is reported for the first time in the international literature.

Rooting at high rates in auxin-free MS medium has also been demonstrated in other perennial Mediterranean shrubs and herbs, such as *A. barba-jovis* [46], *L. monopetalum* [44], *Thymus satureioides* [59], *Lavandula officinalis* [60], *C. nepeta* [36], *C. cretica* [52], *Origanum scabrum* [61], *O. ehrenbergii* and *O. syriacum* [62], and *S. officinalis* and *S. tomentosa* [25], while others such as *T. capitatum* [22], *Lithodora zahnii* [40], and *S. fruticosa* and *S. ringens* [25] required the presence of auxin to exhibit high rooting rates. Further, other Mediterranean xerophytic shrubs such as *A. hermanniae* [24], *S. athoa* [37], *G. alypum* [45], *E. sibthorpii* [41], and *Thymelaea tartonraira* spp. *tartonraira* [63] rooted at low rates in vitro, while *T. hirsuta* failed to root [64,65], confirming the importance of the genome in the in vitro response.

In the present work, the addition of IBA to the medium at 1.0–2.0 mg L^{-1} or 0.5–2.0 mg L^{-1} when shoot-tip or single-node microcuttings, respectively, were placed for rooting, in addition to increasing the number and length of roots, induced intense sprouting of axillary shoots, producing a branched plantlet that did not require pruning after its ex vitro establishment to obtain a marketable shape. Sprouting of axillary shoots during the rooting stage when increased IBA concentration was used, but to a much lesser extent, was found in *A. hermaniae* [24] and *C. cretica* [52].

Microshoot clusters, like microcuttings, rooted 100% in all treatments, including the one without IBA, showing an increase in the number of roots produced per cluster proportional to the increase in IBA concentration, up to the highest IBA concentration tested, i.e., 4 mg L^{-1}, in contrast to microcuttings, which at the highest IBA concentrations showed a reduction in the number of roots. Regarding root length, it was highest at the lowest IBA concentration and decreased slightly, proportionally, with increasing IBA concentration. A negative relation between root number and root length has been shown also in other Mediterranean xerophytes such as *T. capitatum* [22], *L. monopetalum* [44], *D. cruentus* [23], *S. ringens*, and *S. tomentosa* [25]. This could be attributed to the fact that in substrates with higher IBA concentration, we had continuous root production and therefore roots formed close to the date of data collection had a shorter length, thus negatively affecting the average length of roots per microcutting. Alternatively, a higher auxin concentration in the initial stage of rooting favors root induction, but subsequently, in the stage of root elongation, it is an inhibitory factor [22,66].

4.4. Ex Vitro Acclimatization

Ex vitro acclimatization is the final but often most hazardous step in a micropropagation system, as plantlets have to correct abnormal morphology, anatomy, and physiology imposed by in vitro conditions [67]. *B. acetabulosa* plantlets produced by rooting either apical or basal microcuttings, as well as microshoot clusters, were acclimatized and established ex vitro at 100%, presenting morphological characteristics similar to the mother plants. Similarly, the Mediterranean xerophytic shrubs *A. hermaniae* [24], *A. barba-jovis* [67], *L. monopetalum* [44], and *T. capitatum* [22] have been shown to acclimatize at 100% ex vitro, while *Salvia* spp. [25], *C. cretica* [52], *C. nepeta* [36], and *D. fruticosus* [38] also acclimatized at quite high rates, reaching 70–100%. On the contrary, the acclimatization of others, such as *G. alypum*, was very low [45] or failed completely, as in *E. sibthorpii* [41].

5. Conclusions

In this study, *B. acetabulosa*, the Greek horehound, was successfully micropropagated using shoot-tip and single-node explants derived from in vitro-grown seedlings. Both at the establishment stage of the culture and the five subcultures that took place, the explants responded at very high percentages, producing high numbers of shoots on solid MS medium supplemented with 1.0 mg L^{-1} ZEA or BA. The agar concentration is recommended to be increased to 12 g L^{-1} in order to address hyperhydricity problems while maintaining high proliferation indices (92% explant response, 10.2 shoots per explant).

Microcuttings from either the apical part, including the apical meristem, or the basal part of the microshoots, as well as microshoot clusters, rooted 100% on full- or half-strength MS medium, respectively, even in the absence of auxin in the medium. However, the addition of IBA at 1 mg L^{-1} to the rooting medium is recommended, as it induces a greater number of roots and longer roots, as well sprouting of a number of axillary buds in the microcuttings, leading to a suitable plant shape for the floricultural market, without the need for further pruning in the nursery. To our knowledge, this is the first report of the use of microcuttings without the shoot tip for rooting and in vitro plant production.

Plantlets were acclimatized to 100% in ex vitro conditions in 1:1 (v/v) peat/perlite substrate.

Thus, efficient micropropagation protocols for *B. acetabulosa* were developed that allow for further exploitation of the species in the floricultural and pharmaceutical industries. In particular, its use as a landscape plant for arid and semi-arid areas is fully in line with the requirement to enhance vegetation with species that have low water needs, which is becoming even more intense due to the climate crisis we are facing.

Supplementary Materials: The following supporting information can be downloaded at: https://www.mdpi.com/article/10.3390/horticulturae11040390/s1, Table S1: Effect of the experimental factors, i.e., cytokinin type (ZEA, BA, KIN, 2iP) and concentration (0.5, 1.0, 2.0 mg L^{-1}) on the response of *Ballota acetabulosa* single-node explants at the establishment stage; Table S2: Effect of the experimental factors, i.e., explant type (shoot tip, single node), cytokinin type (ZEA, BA, KIN, 2iP) and concentration (0.5, 1.0, 2.0 mg L^{-1}) on the response of *Ballota acetabulosa* explants at the first subculture; Table S3: First subculture: effect of the experimental factors, i.e., cytokinin type (ZEA, BA, KIN, 2iP) and concentration (0.5, 1.0, 2.0 mg L^{-1}) on the response of *Ballota acetabulosa* shoot-tip explants derived from microshoots produced at the establishement stage; Table S4: Effect of the experimental factors, i.e., cytokinin type (ZEA, BA, KIN, 2iP) and concentration (0.5, 1.0, 2.0 mg L^{-1}) on the response of *Ballota acetabulosa* single-node explants at the first subculture; Table S5: Effect of the experimental factors, i.e., subculture number (1st, 2nd, 3rd, 4th) and cytokinin type (ZEA, BA) at 1 mg L^{-1} on the response of *Ballota acetabulosa* shoot-tip explants at the multiplication stage; Table S6: Effect of the experimental factors, i.e., subculture number (1st, 2nd, 3rd, 4th) and cytokinin type (ZEA, BA) at 1 mg L^{-1} on the response of *Ballota acetabulosa* single-node explants at the multiplication stage.

Author Contributions: Conceptualization, M.P. and G.V.; methodology, G.V. and M.P.; validation, M.P. and G.V.; investigation, G.V. and M.P.; resources, M.P.; data curation, G.V.; writing—original draft preparation, G.V. and M.P.; writing—review and editing, M.P. and G.V.; visualization, G.V.; supervision, M.P.; project administration, M.P.; funding acquisition, M.P. All authors have read and agreed to the published version of the manuscript.

Funding: This research was co-financed by the European Regional Development Fund of the European Union and Greek national funds through the Operational Program Education and Lifelong Learning, NRSF 2007-2013. (Project code: MIS 380237; Project: Thalis-ARCHEOSCAPE).

Data Availability Statement: The original contributions presented in this study are included in the article/Supplementary Material. Further inquiries can be directed to the corresponding author.

Conflicts of Interest: The authors declare no conflicts of interest. The funders had no role in the design of the study; in the collection, analyses, or interpretation of data; in the writing of the manuscript, or in the decision to publish the results.

References

1. Dimopoulos, P.; Raus, T.; Bergmeier, E.; Constantinidis, T.; Iatrou, G.; Kokkini, S.; Strid, S.; Tzanoudakis, D. *Vascular Plants of Greece: An Annotated Checklist*; Botanischer Garten und Botanisches Museum Berlin-Dahlem; Hellenic Botanical Society: Berlin, Germany, 2013; p. 106.
2. Davis, P.H. *Flora of Turkey and the East Aegean Islands*; Edinburgh University, Edinburgh Press: Edinburgh, UK, 1982; Volume 7, pp. 156–160.
3. Huxley, A.; Taylor, W. *Flowers of Greece and the Aegean*; Chatto & Windus: London, UK, 1977.
4. Psaras, G.K.; Rhizopoulou, S. Mesophyll structure during leaf development in *Ballota acetabulosa*. *New Phytol.* **1995**, *131*, 303–309. [CrossRef]
5. Sahpaz, S.; Skaltsounis, A.L.; Bailleul, F. Polyphenols from *Ballota acetabulosa*. *Biochem. Syst. Ecol.* **2002**, *30*, 601–604. [CrossRef]
6. Petanidou, T.; Vokou, D. Pollination Ecology of *Labiatae* in a Phryganic (East Mediterranean) Ecosystem. *Am. J. Bot.* **1993**, *80*, 892–899. [CrossRef]
7. Mulas, M. Traditional Uses of *Labiatae* in the Mediterranean Area. *Acta Hortic.* **2006**, *723*, 25–32. [CrossRef]
8. Dulger, B.; Sener, A. Evaluation of antimicrobial activity of *Ballota acetabulosa*. *Afr. J. Microbiol. Res.* **2010**, *4*, 1235–1238.
9. Newall, C.A.; Anderson, L.A.; Philipson, J.D. *Herbal Medicines, a Guide for Health-Care Professionals*; The Pharmaceutical Press: London, UK, 1996; pp. 164–172.
10. Citoglu, G.; Tanker, M.; Sever, B.; Englert, J.; Anton, R.; Altanlar, N. Antibacterial activities of diterpenoids isolated from *Ballota saxatilis* subsp. *saxatilis*. *Planta Med.* **1998**, *64*, 484–485. [CrossRef]
11. Couladis, M.; Tzakou, O.; Verykokidou, E.; Harvala, C. Screening of some Greek aromatic plants for antioxidant activity. *Phytother. Res.* **2003**, *17*, 194–195. [CrossRef] [PubMed]
12. Papafotiou, M.; Bertsouklis, K.; Martini, A.N.; Vlachou, G.; Akoumianaki-Ioannidou, A.; Kanellou, E.; Kartsonas, E. Evaluation of the establishment of native Mediterranean plant species suggested for landscape enhancement in archaeological sites of Greece. *Acta Hortic.* **2017**, *1189*, 177–180. [CrossRef]
13. Egerer, M.; Cecala, J.; Cohen, H. Wild Bee Conservation within Urban Gardens and Nurseries: Effects of Local and Landscape Management. *Sustainability* **2019**, *12*, 293. [CrossRef]
14. Vlachou, G.; Papafotiou, M.; Bertsouklis, K.F. In Vitro Propagation of *Ballota acetabulosa*. *Acta Hortic.* **2016**, *1113*, 171–174. [CrossRef]
15. Chakrabarty, D.; Park, S.Y.; Ali, M.B.; Shin, K.S.; Paek, K.Y. Hyperhydricity in apple: Ultrastuctural and physiological aspects. *Tree Physiol.* **2006**, *26*, 377–388. [CrossRef] [PubMed]
16. Ziv, M. Vitrification: Morphological and physiological disorders of in vitro plants. In *Micropropagation: Technology and Application*; Debergh, P.C., Zimmerman, R.H., Eds.; Kluwer Academic Publishers: Dordrecht, The Netherlands, 1991; pp. 45–69. [CrossRef]
17. Hazarika, B.N. Review. Morpho-physiological disorders in in vitro culture of plants. *Sci. Hortic.* **2006**, *108*, 105–120. [CrossRef]
18. Ivanova, M.; Van Staden, J. Influence of gelling agent and cytokinins on the control of hyperhydricity in *Aloe polyphylla*. *Plant Cell Tissue Organ Cult.* **2011**, *104*, 13–21. [CrossRef]
19. Ivanova, M.; Van Staden, J. Effect of ammonium ions and cytokinins on hyperhydricity and multiplication rate of in vitro regenerated shoots of *Aloe polyphylla*. *Plant Cell Tissue Organ Cult.* **2008**, *92*, 227–231. [CrossRef]
20. Ivanova, M.; Van Staden, J. Natural ventilation effectively reduces hyperhydricity in shoot cultures of *Aloe polyphylla* Schönland ex Pillans. *Plant Growth Regul.* **2010**, *60*, 143–150. [CrossRef]

21. Papafotiou, M.; Bertsouklis, K.F.; Trigka, M. Micropropagation of *Arbutus unedo*, *A. andrachne*, and their natural hybrid, *A. x andrachnoides* from seedling explants. *J. Hortic. Sci. Biotechnol.* **2013**, *88*, 768–775. [CrossRef]
22. Papafotiou, M.; Martini, A.N. In vitro seed and clonal propagation of the Mediterranean aromatic and medicinal plant *Teucrium capitatum*. *HortScience* **2016**, *51*, 403–411. [CrossRef]
23. Bazanis, A.E.; Papafotiou, M. In Vitro Germination, Micropropagation and Addressing the Hyperhydricity of the Balkan Native *Dianthus cruentus*, a Plant with High Ornamental and Xeriscaping Potential. *Horticulturae* **2024**, *10*, 813. [CrossRef]
24. Martini, A.N.; Papafotiou, M. In Vitro Seed and Clonal Propagation of the Mediterranean Bee Friendly Plant *Anthyllis hermanniae* L. *Sustainability* **2023**, *15*, 4025. [CrossRef]
25. Papafotiou, M.; Vlachou, G.; Martini, A.N. Investigation of the Effects of the Explant Type and Different Plant Growth Regulators on Micropropagation of Five Mediterranean *Salvia* spp. Native to Greece. *Horticulturae* **2023**, *9*, 96. [CrossRef]
26. Sarasan, V.; Kite, G.C.; Sileshi, G.W.; Stevenson, P.C. Applications of phytochemical and in vitro techniques for reducing over-harvesting of medicinal and pesticidal plants and generating income for the rural poor. *Plant Cell Rep.* **2011**, *30*, 1163–1172. [CrossRef] [PubMed]
27. Murashige, T.; Skoog, F. A revised medium for rapid growth and bioassays with tobacco tissue cultures. *Physiol. Plant.* **1962**, *15*, 473–497. [CrossRef]
28. Çetin, N.; Mansuroğlu, S.; Önaç, A.K. Xeriscaping Feasibility as an Urban Adaptation Method for Global Warming: A Case Study from Turkey. *Pol. J. Environ. Stud.* **2018**, *27*, 1009–1018. [CrossRef]
29. Caneva, G.; Kumbaric, A.; Savo, V.; Casalini, R. Ecological Approach in Selecting Extensive Green Roof Plants: A Data-Set of Mediterranean Plants. *Plant Biosyst.* **2013**, *149*, 374–383. [CrossRef]
30. Ondoño, S.; Martínez-Sánchez, J.J.; Moreno, J.L. Evaluating the growth of several Mediterranean endemic species in artificial substrates: Are these species suitable for their future use in green roofs? *Ecol. Eng.* **2015**, *81*, 405–417. [CrossRef]
31. Papafotiou, M.; Martini, A.N.; Tassoula, L.; Stylias, E.G.; Kalantzis, A.; Dariotis, E. Acclimatization of Mediterranean Native Sages (*Salvia* spp.) and Interspecific Hybrids in an Urban Green Roof under Regular and Reduced Irrigation. *Sustainability* **2022**, *14*, 4978. [CrossRef]
32. Azeñas, V.; Janner, I.; Medrano, H.; Gulías, J. Performance Evaluation of Five Mediterranean Species to Optimize Ecosystem Services of Green Roofs under Water-Limited Conditions. *J. Environ. Manag.* **2018**, *212*, 236–247. [CrossRef]
33. Tassoula, L.; Papafotiou, M.; Liakopoulos, G.; Kargas, G. Water use efficiency, growth and anatomic-physiological parameters of Mediterranean xerophytes as affected by substrate and irrigation on a green roof. *Not. Bot. Hortic. Agrobot. Cluj-Napoca* **2021**, *49*, 12283. [CrossRef]
34. Kumar, M.; Prasad, Y.; Yadav, A.; Kumar, A. Effects of two different surface sterilization (Sodium Hypochlorite and Mercuric Chloride) agents under in-vitro leaf explant in Gerbera (*Gerbera jamesonii* Bolus). *J. Pharm. Innov.* **2021**, *10*, 1346–1349. Available online: https://www.researchgate.net/publication/361902680 (accessed on 20 January 2025).
35. Martini, A.N.; Vlachou, G.; Papafotiou, M. Effect of Explant Origin and Medium Plant Growth Regulators on In Vitro Shoot Proliferation and Rooting of *Salvia tomentosa*, a Native Sage of the Northeastern Mediterranean Basin. *Agronomy* **2022**, *12*, 1889. [CrossRef]
36. Vlachou, G.; Papafotiou, M.; Bertsouklis, K.F. Studies on Seed Germination and Micropropagation of *Clinopodium nepeta*: A Medicinal and Aromatic Plant. *HortScience* **2019**, *54*, 1558–1564. [CrossRef]
37. Papafotiou, M.; Kalantzis, A. Seed germination and in vitro propagation of *Sideritis athoa*. *Acta Hortic.* **2009**, *813*, 471–476. [CrossRef]
38. Papafotiou, M.; Stragas, J. Seed Germination and in Vitro Propagation of *Dianthus fruticosus*. *Acta Hortic.* **2009**, *813*, 481–484. [CrossRef]
39. Ivanovic, S.; Markovic, M.; Milutinovic, M.; Skocajic, D.; Dunisijevic-Bojovic, D. In Vitro Propagation of *Dianthus cruentus* and Acclimatization in Hydroponic Culture. *Phyton* **2023**, *62–63*, 107–114.
40. Papafotiou, M.; Kalantzis, A. Studies on in vitro propagation of *Lithodora zahnii*. *Acta Hortic.* **2009**, *813*, 465–470. [CrossRef]
41. Bertsouklis, K.F.; Vazaka-Vodena, D.; Bazanis, A.E.; Papafotiou, M. Studies on Seed Germination and Micropropagation of *Ebenus sibthorpii*, an Endemic Shrub of Greece with Potential Ornamental Use. *Horticulturae* **2023**, *9*, 1300. [CrossRef]
42. Máthé, A.; Hassan, F.; Abdul Kader, A. In vitro micropropagation of medicinal and aromatic plants. In *Medicinal and Aromatic Plants of the World*; Máthé, A., Ed.; Springer Science+Business Media: Dordrecht, The Netherlands, 2015; Volume 1, pp. 305–336. [CrossRef]
43. Martini, A.N.; Papafotiou, M. In Vitro Propagation and NaCl Tolerance of the Multipurpose Medicinal Halophyte *Limoniastrum monopetalum*. *HortScience* **2020**, *55*, 436–443. [CrossRef]
44. Bertsouklis, K.; Papafotiou, M.; Balotis, G. Effect of medium on in vitro growth and ex vitro establishment of *Globularia alypum*. *Acta Hortic.* **2003**, *616*, 177–180. [CrossRef]
45. Trigka, M.; Papafotiou, M. In Vitro Propagation of *Anthyllis barba-Jovis* from Seedling Tissues. *Acta Hortic.* **2017**, *1189*, 473–748. [CrossRef]

46. Sreelekshmi, R.; Siril, E.A.; Muthukrishnan, S. Role of Biogenic Silver Nanoparticles on Hyperhydricity Reversion in *Dianthu chinensis* L. An in Vitro Model Culture. *J. Plant Growth Regul.* **2021**, *41*, 23–39. [CrossRef]
47. Mohamed, S.M.; El-Mahrouk, M.E.; El-Banna, A.N.; Hafez, Y.M.; El-Ramady, H.; Abdalla, N.; Dobránszki, J. Optimizing Medium Composition and Environmental Culture Condition Enhances Antioxidant Enzymes, Recovers *Gypsophila paniculata* L Hyperhydric Shoots and Improves Rooting In Vitro. *Plants* **2023**, *12*, 306. [CrossRef] [PubMed]
48. Ivanova, M.; Van Staden, J. Nitrogen source, concentration, and NH_4^+:NO_3^- ratio influence shoot regeneration and hyperhydricity in tissue cultured *Aloe polyphylla*. *Plant Cell Tissue Organ Cult.* **2009**, *99*, 167–174. [CrossRef]
49. Ravanfar, S.A.; Salim, S.; Aziz, M.A.; Abdullah, S.N.A.; Rashid, A.A. Influence of phenyl-urea and adenine-type cytokinins on direct adventitious shoot regeneration of cabbage (*Brassica oleracea* subsp. *capitata*) "KCross". *Plant Biotechnol.* **2014**, *31*, 275–280. [CrossRef]
50. Polivanova, O.B.; Bedarev, V.A. Hyperhydricity in Plant Tissue Culture. *Plants* **2022**, *11*, 3313. [CrossRef] [PubMed]
51. Vlachou, G.; Papafotiou, M.; Bertsouklis, K. Seed germination, micropropagation from adult and juvenile origin explants and address of hyperhydricity of the Cretan endemic herb *Calamintha cretica*. *Not. Bot. Horti Agrobot. Cluj-Napoca* **2020**, *48*, 1504–1518. [CrossRef]
52. Zunazri, N.H.; Kemat, N.; Ariffin, N.; Rineksane, I.A. Effect of media components on hyperhydricity in horticultural crops: A review. *J. Plant Biotechnol.* **2024**, *51*, 307–319. [CrossRef]
53. Casanova, E.; Moysset, L.; Trillas, M.I. Effects of Agar Concentration and Vessel Closure on the Organogenesis and Hyperhydricity of Adventitious Carnation Shoots. *Biol. Plant.* **2008**, *52*, 1–8. [CrossRef]
54. Gerszberg, A. Tissue Culture and Genetic Transformation of Cabbage (*Brassica oleracea* var. *capitata*): An Overview. *Planta* **2018**, *248*, 1037–1048. [CrossRef]
55. Debergh, P.; Aitken-Christie, J.; Cohen, D.; Grout, B.; von Arnold, S.; Zimmerman, R.; Ziv, M. Reconsideration of the term vitrification as used in micropropagation. *Plant Cell Tissue Organ Cult.* **1992**, *30*, 135–140. [CrossRef]
56. Monteuuis, O. In vitro rooting of juvenile and mature *Acacia mangium* microcuttings with reference to leaf morphology as a phase change marker. *Trees* **2004**, *18*, 77–82. [CrossRef]
57. Preece, J. Stock Plant Physiological Factors Affecting Growth and Morphogenesis. In *Plant Propagation by Tissue Culture: Volume 1. The Background*, 3rd ed.; Edwin, F.G., Michael, A.H., de Klerk, G.-J., Eds.; Springer: Dordrecht, The Netherlands, 2008; pp. 403–422.
58. Aicha, N.; Rachida, T.C.; El Abdelmalek, M. Micropropagation of *Thymus satureioides* Coss. an endangered medicinal plant of Morocco. *J. Agric. Technol.* **2013**, *9*, 487–501.
59. Tyub, S.; Kamili, A.N.; Shah, A.M. Effect of BAP on Shoot Regeneration in Shoot Tip Cultures of *Lavandula officinalis*. *J. Res. Dev.* **2007**, *7*, 125–130.
60. Alexopoulos, A.A.; Kartsonas, E.; Karras, S.; Mavrommati, E.; Petropoulos, S.A.; Papafotiou, M. In Vitro Propagation of *Origanum scabrum* (Boiss. & Heldr.): An Endemic Medicinal Plant of Greece. *Plants* **2023**, *12*, 2118. [CrossRef]
61. El Beyrouthy, M.; Elian, G.; Abou Jaoudeh, C.; Chalak, L. In vitro propagation of *Origanum syriacum* and *Origanum ehrenbergii*. *Acta Hortic.* **2015**, *1083*, 169–172. [CrossRef]
62. Martini, A.N.; Papafotiou, M. Effect of cytokinins on in vitro blastogenesis of *Thymelaea tartonraira* ssp. *tartonraira* (L.) All. *Acta Hortic.* **2019**, *1242*, 511–514. [CrossRef]
63. Papafotiou, M. In vitro propagation of temperate zone woody plants with potential ornamental use. *Acta Hortic.* **2010**, *885*, 255–262. [CrossRef]
64. Martini, A.N.; Papafotiou, M. Investigation of micropropagation of the Mediterranean xerophyte *Thymelaea hirsuta* (L.) Endl. (*Thymelaeaceae*). *Acta Hortic.* **2020**, *1298*, 335–340. [CrossRef]
65. De Klerk, G.J.; Der Krieken, W.V.; De Jong, J.C. Review the formation of adventitious roots: New concepts, new possibilities. *Vitro Cell. Dev. Biol. Plant* **1999**, *35*, 189–199. [CrossRef]
66. Pospíšilová, J.; Tichá, I.; Kadleček, P.; Haisel, D.; Plzáková, Š. Acclimatization of micropropagated plants to ex vitro conditions. *Biol. Plant.* **1999**, *42*, 481–497.
67. Vlachou, G.; Papafotiou, M.; Bertsouklis, K.F. Studies on in vitro propagation of *Anthyllis barba-jovis* L. *Acta Hortic.* **2017**, *1155*, 317–320. [CrossRef]

horticulturae

MDPI

Article

Effect of Cutting Phenological Stage, Chemical Treatments, and Substrate on Rooting Softwood Cuttings of Tree Peony

Dongli Li [1], Fangyun Cheng [1,*], Xiwen Tao [2] and Yuan Zhong [1]

1 State Key Laboratory of Efficient Production of Forest Resources, Beijing Key Laboratory of Ornamental Plants Germplasm Innovation & Molecular Breeding, National Engineering Research Center for Floriculture, Peony International Institute, School of Landscape Architecture, Beijing Forestry University, Beijing 100083, China; 13269890293@163.com (D.L.); zhongyuanbjfu@126.com (Y.Z.)

2 Beijing Guose Peony Technology Co., Ltd., Beijing 102100, China; 13161186033@163.com

* Correspondence: chengfy8@263.net; Tel.: +86-10-62338304

Abstract: Tree peony is well-known for its ornamental value, medicine function, oil, and edible use. However, the difficulty in propagating tree peony impedes its cultivation and industrial advancement. Softwood cutting is an effective method to promote the propagation of tree peony. This research investigated the effects of several factors (cultivar, cutting phenological stage, auxin type, polyamine, and substrate) on the rooting of softwood cutting in tree peony. The results showed that rooting ability varied with cultivars and cutting phenological stages, with the highest rooting rates being for 'High noon' and 'Jinghua Qingxue' during the vigorous growth stage, reaching 50% and 53.33%, respectively. IBA 2000 mg·L^{-1} was optimal for rooting in 'High noon' cuttings, with the maximum root number (5.67) and root length (6.3 cm). Putrescine of 1.0 mM could significantly improve the rooting rate and rooting quality of 'Jinghua Qingxue' cuttings, which had the highest rooting rate of 54.17% in the cocopeat/perlite substrate (*v:v* 1:1). Anatomical observation showed that most adventitious roots were generated from callus meristem nodules differentiated from cortical parenchyma cells while a few came from stem bark, as well as integrated root induction. This study is an innovation in and supplement to tree peony propagation research, and a propagation protocol was primarily established for softwood cuttings in tree peony.

Keywords: phenological stage; plant growth regulator; polyamine; rooting; softwood cutting; substrate; tree peony

1. Introduction

Tree peony (*Paeonia* sect. Moutan) is a traditional ornamental flower which is also used in medicine, as an oil, and in cosmetology and food science [1,2]. However, the difficulty in propagating tree peony impedes its cultivation and industrial advancement [3]. Softwood cuttings have been widely used in the industrial production of commercial seedlings for many ornamental and crop plants, due to their advantages of a high reproduction coefficient, short growth cycle, and simple operation [4]. Meanwhile, softwood cutting is a potent supplementary method to promote difficult-to-root species propagation and early flowering and fruiting in many plants [5–7].

The success of softwood cuttings depends on the formation of adventitious roots (ARs), which is influenced by both endogenous and exogenous factors [8]. Many factors, such as genotypes, phenological stage, plant growth regulators (PGRs), polyamines, and substrates, have been confirmed to be involved in the rooting of cuttings in various plants [9–12]. The rooting ability of cuttings is closely related to genotypes, phenological stage, and climatic conditions which lead to seasonal variation in woody plants [13,14]. Johnson et al. [10] investigated 20 wild and cultivated *Prunus* genotypes and found that the plum species rooted at a higher percentage and produced more roots than the peach species, while the almond species generally failed to root. The rooting percentage varied from 0% to 62.5% in diverse cultivars of *Pistacia lentiscus* L. [15] and exhibited significant differences in seven pomegranate cultivars [7]. As reported, cutting specifications also have an important influence on the rooting of cuttings, including the length, diameter, and the amounts of leaves [5,16–18]. Furthermore, many studies have shown that the phenological stages of the stock plants play an important role in the propagation of softwood cuttings, indicating that the homeostasis between root-promoting substances and root-inhibiting substances in the stock plants changes with seasonal variation [14,16]. It was reported that four difficult-to-root rose cultivars (Duchesse d'Angoulême, Hurdals, Maiden's Blush, and Mousseuse Rouge) showed different rooting abilities at four phenological stages and that all cultivars had the highest rooting percentage at the closed flower buds stage, while none of the cuttings could root during the 10~14 days after petal shedding [5].

Among these exogenous rooting factors, indole-3-butyric acid (IBA) is widely used as an efficient stimulant to promote the rooting of cuttings in many woody plants, including *Rosa damascena* [19], *Sterculia foetida* [20], *Prunus cerasifera* [10], *Acer rubrum* [21], *Photinia x fraseri* [22], and *Juglans regia* [23]. The effect of exogenous IBA application on rooting depends on the rooting capacity of various genotypes [21,24]. Moreover, the role of IBA is supposed to convert into indole-3-acetic acid (IAA) via β-oxidation in peroxisome [25], by which IBA can upregulate the auxin signaling pathway and enhance the accumulation of IAA at the rooting sites to promote rooting [23]. In some cases, IBA alone could not promote the formation of AR in cuttings, while the combination of IBA and NAA resulted in better rooting rate [26,27].

Polyamines (PAs) alone or combined with IBA are beneficial for rooting in many plants, including *Corylus avellana* [28], *Olea europaea* [29], *Prunus persica x Prunus amygdalus* [30], and *Rosa damascena* [12]. As reported, putrescine (Put), spermidine (Spd) and spermine (Spm) had positive impacts on AR formation in *Rosa damascena* (difficult-to-root) and Put led to a better rooting performance when used in conjunction with IBA [12]. PAs have been proposed as rooting markers since their endogenous accumulation has been positively correlated with the rooting induction and initiation phases [30,31].

In addition, the type of substrates have a significant effect on rooting parameters in stem cuttings and might be genotype-dependent [32–34]. The highest rooting rate of *Pinus massoniana* cuttings occurred in sand (75.17%), followed by perlite (60.52%), while the rooting parameters, including root number, root length, root diameter and rooting effect index, were superior in perlite than other substrates [11]. Recently, some soilless substrates have been widely used for commercial cuttings, including rockwool, horticultural foam, and aeroponics [33,35,36], among which aeroponics was more effective than other methods due to there being less algae or mold over time [34].

However, few studies have been reported on softwood cutting in tree peony. The test of leaf-bud cuttings was carried out in four cultivars of tree peony and two of them successfully rooted from the base callus two months after the cutting [37]. The rooting rate of five tree peony cultivars varied from 20% to 91% through etiolated softwood cuttings, with the superior substrate of vermiculite [38]. It was found that large amounts of callus tissues were induced at the incision base before rooting and that ARs finally protruded out of callus surface [39,40]. It is important to determine the key factors influencing the rooting efficiency of tree peony softwood cutting. Accordingly, the aim of the study was to evaluate the effects of phenological stage, auxin types, polyamines, and substrates on the rooting of nine tree peony cultivars to improve nursery propagation, and the optimal cutting propagation protocol was summarized.

2. Materials and Methods

2.1. Plant Materials

In 2021–2022, nine cultivars of tree peony (Itoh peony 'Bartzella', *Paeonia* × *lemoinei* 'High Noon', *Paeonia rockii* 'Jinghua Qingxue', 'Jing Yunguan', 'Jinghong', 'Jing Zuimei', 'Jing Helan', 'Gaoyuan Shenghuo', and *Paeonia suffruticosa* 'Luoyang Hong') were chosen from the plants growing in the nursery of Beijing Guose Peony Garden (40°45′ N, 115°97′ E), Beijing, China. The softwood and semi-lignified cuttings were collected from the current shoots after pruning rejuvenation last fall, then trimmed to 6~8 cm long with 1~2 buds and sterilized with carbendazim for further experiments.

2.2. Cutting Treatments and Experimental Conditions

2.2.1. Cutting Phenological Stage

The cuttings of nine cultivars (refer to Section 2.1) were prepared at different phenological stages from 10 days before flowering (DBF) to 90 days after flowering (DAF). The cuttings were inserted in peat/perlite substrate (*v:v* 6:4) after the basal parts (2 cm) were soaked in IBA solution (2000 mg·L^{-1}) for 15 min. For each collection time and cultivar, 10 cuttings with 3 replicates were considered.

2.2.2. Plant Growth Regulators

The cuttings were taken from the tree peony 'High Noon' at 10 DBF in 2021. The treatment of plant growth regulators was detailed in Table 1. IBA solutions were prepared by dissolving IBA powder (Biorigin, Beijing, China) into distilled water using 1 mol·L^{-1} of sodium hydroxide. Genpan (active ingredient 2.5% IBA + 2.5% NAA, Sichuan Guoguang Agrochemical Co., Ltd., Chengdu, China) solutions were diluted with distilled water to the corresponding ratio, and rooting powder (active ingredient 0.25%IBA, Rhizopon, Hazerswoude, the Netherlands) was directly used, with water as a control. The basal parts (2 cm) of cuttings were dipped into these solutions for 15 min or into rooting powder for 3 s and they were inserted into the peat/perlite substrate (*v:v* 6:4). For each treatment, 10 cuttings with 3 replicates were considered.

Table 1. The treatment of plant growth regulators on rooting in the tree peony 'High Noon'.

Treatment Code	Plant Growth Regulator	Concentration (mg·L^{-1}/Dilution Factor)	Soaking Time (min/s)
IBA-1000	IBA	1000	15
IBA-2000	IBA	2000	15
IBA-3000	IBA	3000	15
GP-150	GenPan	150	15
GP-250	GenPan	250	15
GP-350	GenPan	350	15
RP	Rooting Powder	0.25% IBA	3 s
CK	Water	-	15

2.2.3. Polyamines

The experiment into how polyamines influence rooting was conducted in the tree peony 'Jinghua Qingxue' at 19 DAF in 2022. In this experiment (Table 2), the basal parts (2 cm) of the cuttings were initially dipped in solutions of putrescine, spermidine, and spermine at three concentrations (0.05 mM, 0.1 mM, and 1.0 mM) for 5 mins and then dipped in IBA solution (2000 mg·L^{-1}) for 15 min with water as a control. Finally, the cuttings were inserted into a perlite/vermiculite substrate (*v:v* 7:3). For each treatment, 10 cuttings and 3 replicates were considered.

Table 2. The treatment of polyamines on rooting in the tree peony 'Jinghua Qingxue'.

Treatment Code	Polyamine	Concentration (mM)	Soaking Time (min)
Put-0.05	Put	0.05	5
Put-0.1	Put	0.1	5
Put-1.0	Put	1.0	5
Spd-0.05	Spd	0.05	5
Spd-0.1	Spd	0.1	5
Spd-1.0	Spd	1.0	5
Spm-0.05	Spm	0.05	5
Spm-0.1	Spm	0.1	5
Spm-1.0	Spm	1.0	5
CK	Water	-	5

2.2.4. Substrates

The experiment of substrates influencing rooting was conducted on the tree peony 'Jinghua Qingxue' at 19 DAF in 2022. In this experiment, the cuttings were treated with IBA 2000 mg·L^{-1} for 15 min and inserted in seven rooting substrates: (T1) cocopeat, (T2) cocopeat-perlite (*v:v* 1:1), (T3) cocopeat-perlite (*v:v* 1:2), (T4) perlite-vermiculite (*v:v* 7:3), (T5) perlite-peat (*v:v* 3:1), (T6) perlite-peat (*v:v* 4:6), (T7) perlite-peat-vermiculite (*v:v:v* 3:1:1). For each treatment, 10 cuttings with three replications were considered.

2.2.5. Experimental Conditions

All these experiments were conducted in the greenhouse of the Beijing International Flower Port (40°10′ N, 116°47′ E), Beijing, China. An intelligent controlled fog system maintained the optimal climatic conditions (air temperature 18 °C~30 °C and relative humidity 75~90%). The cuttings were artificially water-sprayed every 2 h to keep the leaves moist during the day and covered with a shading net (shading percentage of 70%) that reduced the light intensity. The cuttings were protected by applying fungicides and pesticide every 7 days during rooting: the dilute solutions (800~1000) consisted of carbendazim (Sichuan Guoguang Agrochemical Co., Ltd., Chengdu, China), dimethomorph (Acrobat-BASF, Shanghai, China), mancozeb (Limin Chemical Co., Ltd., Xinyi, China), and imidacloprid (Sichuan Guoguang Agrochemical Co., Ltd., Chengdu, China).

2.3. *Evaluation of Rooting Cuttings*

Three cuttings were randomly selected for observation every 7 days in each treatment and returned to the substrate for continuous growth. The adventitious roots appeared at the surface of callus or protruded from bark of the stem at approximately 60 days after cutting (DAC). Then, 90 DAC, the rooting parameters were calculated to evaluate the effects of developmental stage, cultivar, plant growth regulators, polyamines, and substrates on the rooting of cuttings, including rooting rate, callus rate, rot rate, average root number, and root length. The calculation formulas were as follows:

$$\text{Rooting rate} = (\text{Number of rooted cuttings}/\text{Total number of cuttings}) \times 100\%$$

$$\text{Callus rate} = (\text{Number of callusing cuttings}/\text{Total number of cuttings}) \times 100\%$$

$$\text{Rot rate} = (\text{Number of rot cuttings}/\text{Total number of cuttings}) \times 100\%$$

$$\text{Average root number} = (\text{Total root number}/\text{Total number of rooted cuttings}) \times 100\%$$

$$\text{Average root length} = (\text{Total root length}/\text{Total root number}) \times 100\%$$

2.4. Morphological and Anatomical Observation on Shoots Development and Rooting Process

To study the developmental stage of shoots and the rooting process of cuttings, the tree peony 'High Noon' was chosen for morphological and anatomical observation. Five shoots were collected and recorded at each of the four developmental stages of 10 DBF, 28 DAF, 55 DAF (2022), and 80 DAF in 2021, respectively. Then, the basal part (2 cm) of the cuttings was cut into pieces (2 mm thickness) and put into FAA fixative (90 mL 70% alcohol, 5 mL 38% formaldehyde, and 5 mL acetic acid), and the anatomical structure was studied through paraffin sectioning [41]. During rooting process, the rooting sites of 'HN' cuttings at about 60 DAC were cleaned carefully and cut into pieces (2 mm thickness) for paraffin sections to identify the origin of adventitious roots [42]. After staining with safranin O-fast green (Biorigin, Beijing, China), the paraffin sections were sealed with neutral gum and photographed under a SOPTOP optical microscope (Leica, Wetzlar, Germany).

2.5. Statistical Analysis

The data analysis was performed using Excel (2019) (Microsoft Corporation, Redmond, WA, USA) and data were expressed as the mean ± standard deviation. The comparisons between different data were analyzed by one-way analysis of variance (ANOVA) ($p < 0.05$) and Duncan's test using SPSS software (version 27.0; IBM, Chicago, IL, USA). The statistical significance level was set at $p < 0.05$.

3. Results

3.1. Effect of Cutting Phenological Stage on the Rooting of Tree Peony

In the 2021 test, the cuttings of four cultivars could root after we prepared them from shoots at the developmental stages of both 10 DBF and 30 DAF, and only a few 'HN' cuttings rooted at 80 DAF (Table 3). The highest rooting rate of 'HN' cuttings was obtained from the 28 DAF plants, reaching 50%, followed by rates of 38.33% and 6.67% obtained from the 10 DBF and 80 DAF plants, respectively. Higher rooting rates for the other three cultivars were obtained from 10 DBF plants, achieving rooting rates of 31.67%, 21.67%, and 16.67% for 'JHQX', 'JYG', and 'GYSH', respectively. It was found that the rooting ability of cuttings was consistent with the callus rate and opposite to the rot rate in the four cultivars (Table 3).

In the 2022 test, the rooting rates varied significantly, ranging from 0 to 53.33% at different developmental stages among the nine cultivars (Table 3). Obviously, higher rooting rates of these cuttings were noticed at around 20~40 DAF, but no rooting was newly generated beyond 60 DAF. Among these cultivars, the highest rooting rate was obtained from 'JHQX' at 19 DAF, achieving 53.33%, followed by a rate of 30% obtained from 'BZL' at 36 DAF (Table 3). Along with the maturity of shoots for softwood cuttings, the rates of rooting and callusing in these cultivars gradually decreased while the rot rates increased (Table 3), indicating that the pivotal role of developmental stages of shoots in the rooting of tree peony softwood cuttings.

Table 3. The effect of cutting phenological stage on rooting in tree peony.

Year	Cultivar	DBF/DAF	Rooting Rate (%)	Callus Rate (%)	Rot Rate (%)
2021	HN	10DBF	38.33 ± 10.27 ab	100.00 a	6.67 ± 4.71 j
		28DAF	50.00 ± 8.16 a	90 ± 8.16 ab	33.33 ± 4.71 g
		80DAF	6.67 ± 4.71 c	40 ± 8.16 e	70 ± 8.16 d
	JHQX	10DBF	31.67 ± 6.24 ab	100.00 a	50 ± 8.16 f
		29DAF	13.33 ± 4.17 bc	20 ± 8.16 h	70 ± 8.16 d
		80DAF	0.00 c	16.67 ± 4.71 hi	73.33 ± 9.43 d
	JYG	10DBF	21.67 ± 8.50	76.67 ± 12.47 c	63.33 ± 4.71 e
		29DAF	6.67 ± 4.71 c	13.33 ± 4.71 j	81.67 ± 6.24 bc
		81DAF	0.00 c	6.67 ± 4.71 j	93.33 ± 4.71 ab
	GYSH	10DBF	16.67 ± 9.43 bc	80 ± 8.16 d	66.67 ± 4.71 de
		30DAF	6.67 ± 4.71 c	13.33 ± 4.71 j	83.33 ± 4.71 bc
		81DAF	0	6.67 ± 4.71 j	96.67 ± 4.71 a
2022	HN	18DAF	16.67 ± 4.71 bc	96.67 ± 4.71 a	10.00 ± 8.16 i
		35DAF	10.00 ± 8.16 c	93.33 ± 4.71 bc	23.33 ± 4.71 h
		55DAF	6.67 ± 4.71 c	40.00 ± 8.16 fg	70.00 ± 8.16 d
		70DAF	0.00 c	21.67 ± 8.50 gh	83.33 ± 4.71 bc
	BZL	36DAF	30.00 ± 8.16 ab	100.00 a	13.33 ± 4.71 i
		58DAF	6.67 ± 4.71 c	80.00 ± 8.16 c	23.33 ± 12.47 h
		72DAF	0.00 c	16.67 ± 9.43 j	80.00 ± 8.16 bc
	JHQX	19DAF	53.33 ± 12.47 a	96.67 ± 4.71 ab	6.67 ± 4.71 j
		36DAF	7.50 ± 5.40 c	85.83 ± 4.25 bc	25.00 ± 4.08 h
		56DAF	3.33 ± 4.71 c	39.17 ± 8.25 fg	71.67 ± 2.36 d
	GYSH	20DAF	6.67 ± 4.71 c	46.67 ± 4.71 e	66.67 ± 9.43 de
		38DAF	3.33 ± 4.71 c	25.00 ± 4.08 gh	75.00 ± 4.08 cd
		58DAF	0.00 c	10.00 ± 8.16 ij	80.00 ± 8.16 bc
	JYG	10DAF	0.00 c	10.00 ± 8.16 ij	86.67 ± 4.71 ab
		22DAF	10.83 ± 1.18 c	35.83 ± 4.25 ef	78.33 ± 2.36 cd
		37DAF	0.00 c	6.67 ± 4.71 j	85.83 ± 4.25 b
	JZM	10DAF	0.00 c	17.50 ± 3.54 hi	78.33 ± 2.36 cd
		45DAF	6.67 ± 4.71 c	29.17 ± 7.17 fg	70.83 ± 7.17 d
		68DAF	0.00 c	3.33 ± 4.71 k	83.33 ± 4.71 bc
	JH	40DAF	3.33 ± 4.71 c	21.67 ± 2.36 hi	85.83 ± 4.25 b
		62DAF	0.00 c	3.33 ± 4.71 k	93.33 ± 4.71 ab
	JHL	41DAF	6.67 ± 4.71 c	20.83 ± 7.16 h	89.17 ± 1.18 b
		63DAF	0.00 c	3.33 ± 4.71 k	96.67 ± 4.71 a
	LYH	44DAF	3.33 ± 4.71 c	20.00 ± 8.16 h	75.00 ± 4.08 d
		65DAF	0.00 c	3.33 ± 4.71 k	93.33 ± 4.71 ab

Note: The abbreviations of the nine cultivars are BZL, HN, JHQX, JYG, JH, JZM, JHL, GYSH, and LYH for 'Bartzella', 'High Noon', 'Jinghua Qingxue', 'Jing Yunguan', 'Jinghong', 'Jing Zuimei', 'Jing Helan', 'Gaoyuan Shenghuo', and 'Luoyang Hong', respectively. The flowering stages were on the 15 May 2021 and 7 May 2022, respectively. DBF: day before flowering stage, DAF: day after flowering stage. The cuttings were dipped in IBA solution (2000 mg·L^{-1}) for 15 min and inserted in peat/perlite substrate (*v:v* 6:4). Different letters in the same column indicate significant differences at the p level below 0.05.

3.2. Effect of Plant Growth Regulators on Rooting

The rooting rates of 'HN' cuttings with all PGR treatments were higher than the control group, as well as the callus rates (Figure 1A). The highest rooting rate was obtained with the use of IBA 2000 mg·L^{-1} (38.33%), followed by 1000 mg·L^{-1} (36.67%). A lower rooting rate was noticed for RP (10%), just higher than the control (5.89%). Furthermore, a callus rate of 100% was achieved with IBA 2000 mg·L^{-1}, GP-250, and GP-350. However, the rot rate varied significantly among all the PGR treatments, with the lowest and highest rot rates obtained with IBA 2000 mg·L^{-1} (6.67%) and GP-150 (60.83%), respectively (Figure 1A). Additionally, PGRs also significantly affected the root number and root length of cuttings. The largest root number occurred in the treatment with RP (8.0), followed by IBA 2000 mg·L^{-1} (5.67) (Figure 1B). The longest root length was obtained with the treatment of GP-150 (6.6 cm), followed by RP (6.47 cm) and IBA 2000 mg·L^{-1} (6.37 cm) (Figure 1C). Considering all of these results, IBA 2000 mg·L^{-1} was the optimal treatment to promote rooting for softwood cuttings in the tree peony 'High noon'.

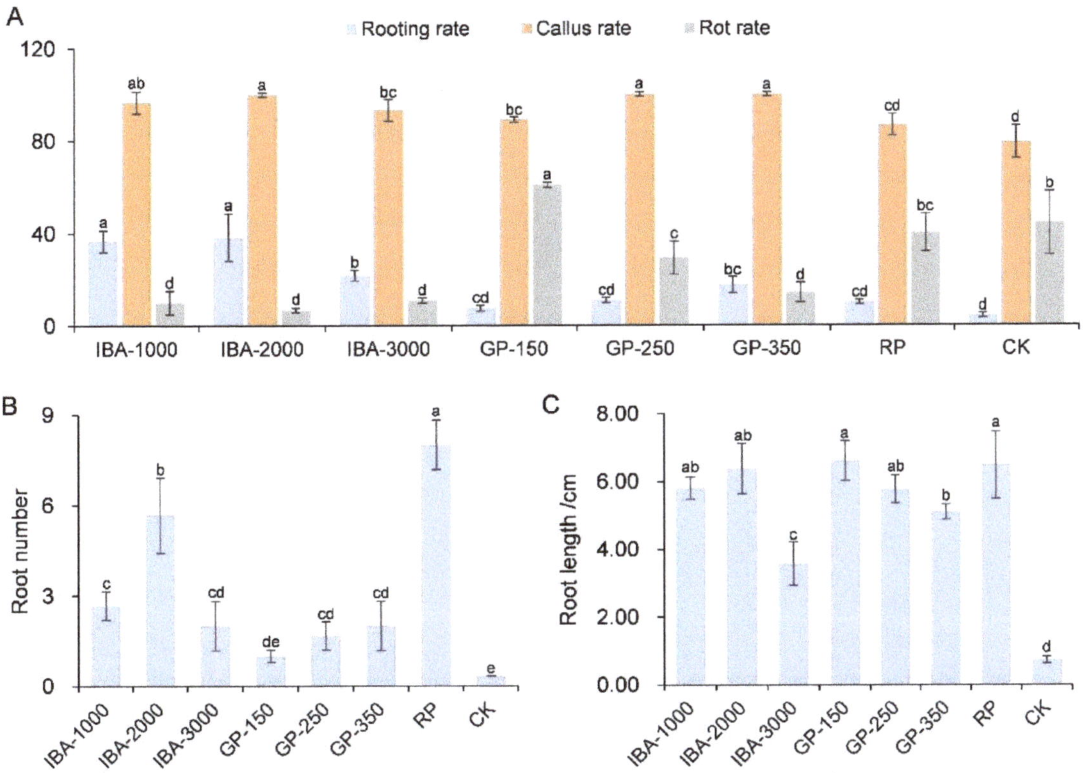

Figure 1. Effect of plant growth regulators on rooting parameters of 'High Noon' cuttings. (**A–C**) represent the rates of rooting/callus/rot, root number, and root length with different plant growth regulators on the rooting of 'High Noon' cuttings, respectively. Different letters indicate significant differences at the *p* level below 0.05.

3.3. Effect of Polyamines on Rooting

Combined with 2000 mg·L^{-1} IBA, the rates of rooting and callusing were higher with Put treatments in 'JHQX' cuttings, and only few cuttings could root in Spd-0.1 and Spm-1.0 treatments, while the cuttings hardly rooted in the remaining treatments (Figure 2A). The highest rooting rate was obtained by Put-1.0 (25%), followed by Put-0.1 (20.83%) and Put-0.05 (16.67%). Meanwhile a higher callus rate of 95.83% was achieved with Put-1.0 and Put-0.05 compared with other treatments. Additionally, lower rot rates ranged from 41.67% to 45.83% in Put, and higher rot rates reached 70.83~91.67% in the treatments of Spd and Spm as well as in the control group. Furthermore, Put-1.0 could produce the maximum root number and root length, reaching 6.33 and 4.0 cm, respectively (Figure 2B,C), which was more conducive to later growth for cuttings. Therefore, Put-1.0 was suitable for promoting AR formation in 'JHQX' softwood cuttings.

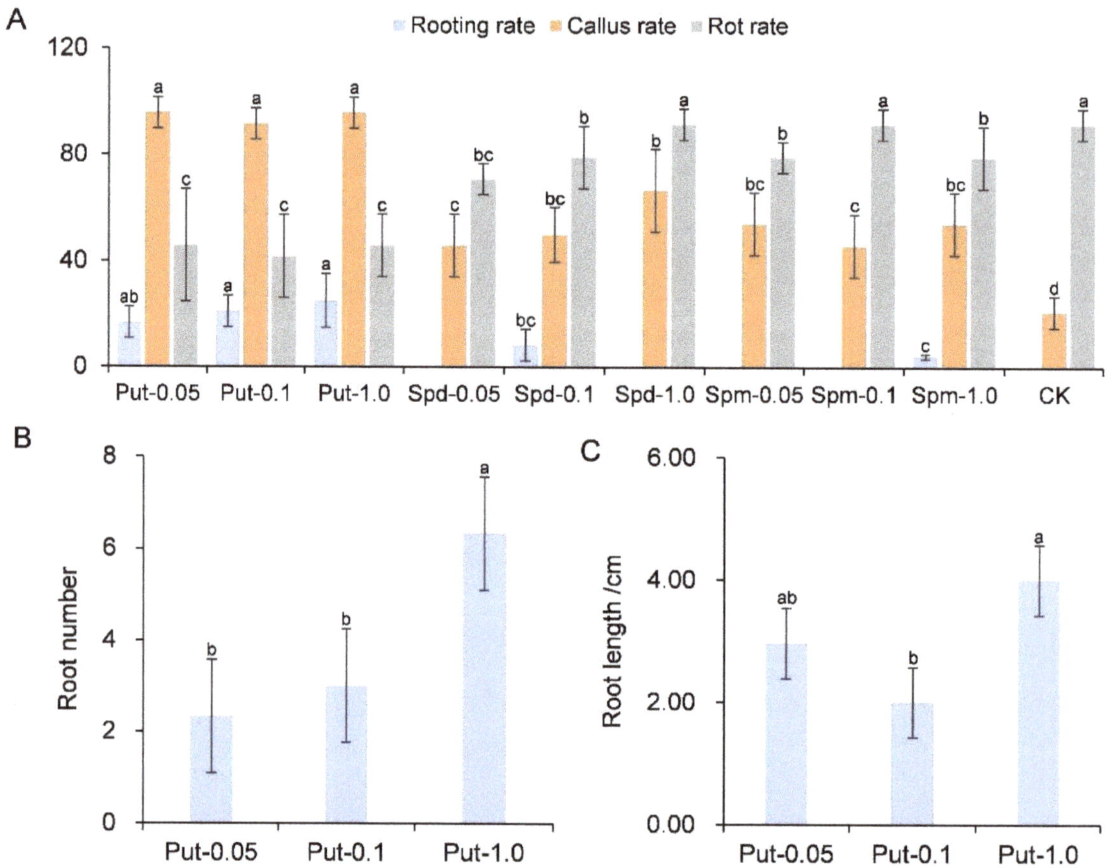

Figure 2. Effect of polyamines on rooting parameters of 'Jinghua Qingxue' cuttings. (**A**–**C**) represent rates of rooting/callus/rot, root number, and root length of different polyamines on the rooting of 'Jinghua Qingxue' cuttings, respectively. The cuttings were initially dipped in the solutions of Put, Spd, and Spm at three concentrations (0.05 mM, 0.1 mM, and 1.0 mM) for five mins and then dipped in IBA solution (2000 mg·L^{-1}) for 15 min with water as a control. Different letters indicate significant differences at the *p* level below 0.05.

3.4. *Effect of Substrates on Rooting*

The rooting rate and quality varied significantly among the seven substrates in 'JHQX' softwood cuttings (Figure 3A–C). Rooting rates obtained from T2, T6, and T4, reaching 54.17%, 53.33%, and 37.5%, respectively, were higher than the rates of 16.67%, 8.33%, and 4.17% obtained from T5, T7, and T1, respectively (Figure 3A), while none of the cuttings could root in substrate T3. Similarly, higher callus rates were obtained from T2, T4, T5, and T6, ranging from 91.67% to 96.67%. Lower callus rates were obtained from other three substrates, ranging from 45.83%~62.5%. Our analysis indicated that the lowest rot rate was obtained from T6 (6.67%), followed by T4 (25%) and T2 (37.5%) (Figure 3A). Furthermore, the maximum root number (8.33) and root length (10.93 cm) were significant in T2, followed by T6 with a root number of 5.33 and a root length of 7.83 cm (Figure 3B,C), indicating that the rooting performance was better in T2 than T6. On the basis of these results, T2 substrate was suitable for callusing and rooting in softwood cuttings of the tree peony 'JHQX'.

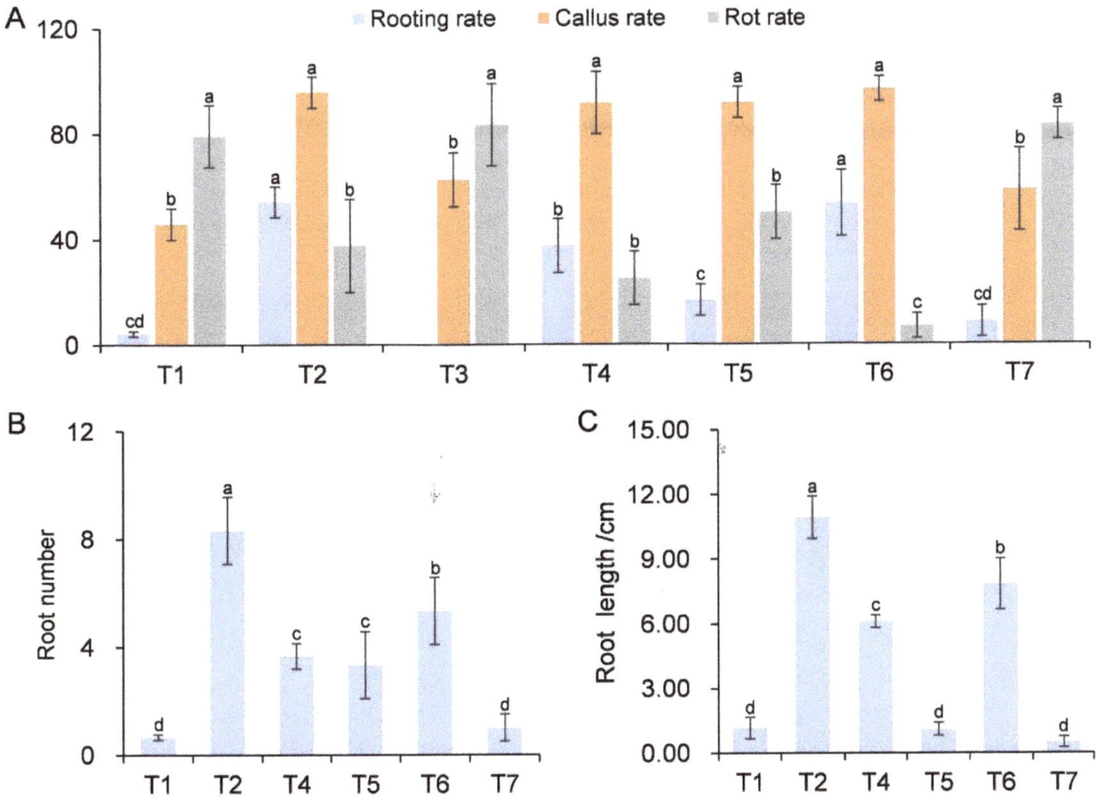

Figure 3. Effect of substrates on rooting parameters of 'Jinghua Qingxue' cuttings. (**A–C**) represent rates of rooting/callus/rot, root number, and root length of different substrates on rooting of 'Jinghua Qingxue' cuttings, respectively. T1, T2, T3, T4, T5, T6, and T7 represent cocopeat (100%), cocopeat/perlite (*v:v* 1:1), cocopeat/perlite (*v:v* 1:2), perlite/vermiculite (*v:v* 7:3), perlite/peat (*v:v* 3:1), perlite/peat (*v:v* 4:6), and perlite/peat/vermiculite (*v:v:v* 3:1:1), respectively. The cuttings were treated with IBA 2000 mg·L^{-1} for 15 min and inserted into the seven substrates. Different letters indicate significant differences at the *p* level below 0.05.

3.5. Softwood Development and Rooting of Softwood Cuttings in Tree Peony

3.5.1. Morphological and Anatomical Observation of 'HN' Softwood Development

Based on the results in Section 3.1, the rooting ability of tree peony cuttings was closely associated with the developmental stages of shoots; thus, the current shoots for 'HN' softwood cuttings were observed in terms of their morphological and anatomical aspects at different developmental stages in 2021 (Figure 4). We found that the shoot stem was initially very tender and red–purple at 10 DBF (Figure 4A), then turned hard and green at 28 DAF (Figure 1B) and gradually became yellow beyond 55 DAF (Figure 4C,D). Synchronously, the leaves of shoots were primarily not yet unfolded and red–purple at 10 DBF (Figure 4A), gradually unfolded with extension and turned green at 28 DAF, then turned yellow beyond 55 DAF (Figure 4B–D), which indicated that the contents of chlorophyll and carotenoid and photosynthetic capacity of shoots varied with developmental stages [43]. Via further observation through an optical microscope, the primary structure of cuttings stem consisted of the epidermis, cortex, phloem, xylem, and pith from the outside to inside at 10 DBF (Figure 4a), indicating that the differentiation of shoots had not yet been completed and that the cells possessed a high potential differentiation capability. The parts of phloem and xylem were found to be obviously differentiated and thickened at 28 DAF (Figure 4b), when the shoots were at the vigorous growth stage. Beyond 55 DAF, a ring of sclerenchyma [44] was noticed outside the phloem and the xylem layer gradually thickened (Figure 4c,d), which indicated that the shoots were gradually undergoing lignification and that the rooting ability might decrease [24]. As mentioned above, the highest rooting rate of 'HN' cuttings was obtained from 28 DAF (50%) and subsequently from 10 DBF (38.33%). Therefore, the developmental stages of shoots played an important role in the adventitious rooting of softwood cutting for tree peony, and the cuttings were more likely to root when prepared from shoots in vigorous growth.

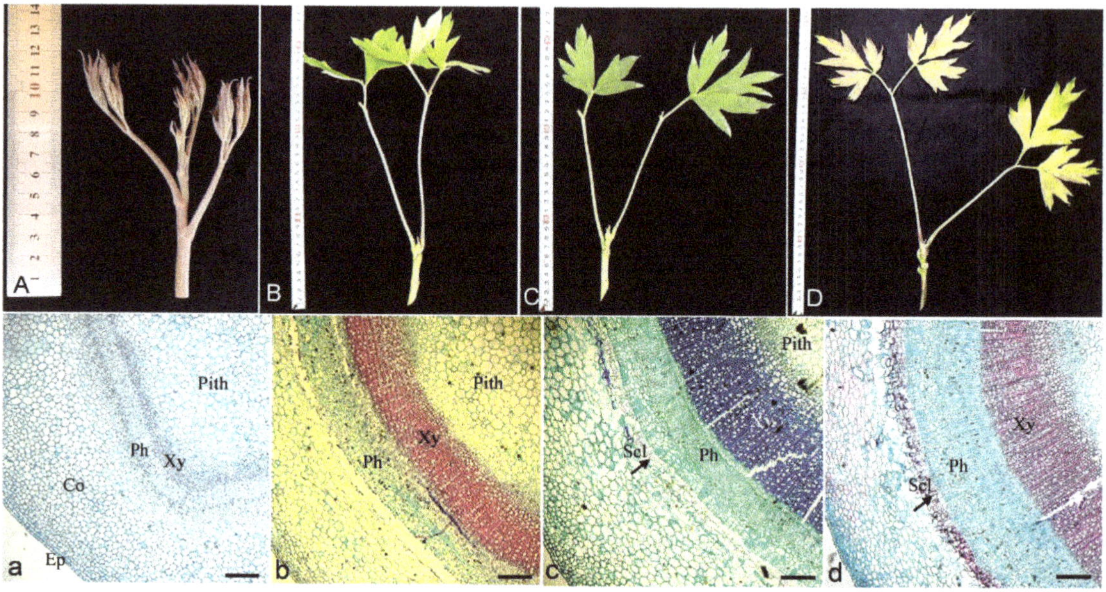

Figure 4. The morphological and anatomical structure of shoot section at different phenological stages in the tree peony 'HN'. (**A–D**) represent the status of cuttings at 10 DBF, 28 DAF, 55 DAF (2022), and 80 DAF in 2021, respectively. DBF: day before flowering stage, DAF: day after flowering stage. (**a–d**) indicated the anatomical structure of the stem from 10 DBF to 80 DAF. The flowering stages were on 15 May 2021 and 7 May 2022, respectively. Abbreviations: Ep, epidermis; Co, cortex; Scl, sclerenchyma; Ph, phloem; Xy, xylem. (**a–d**): bar = 100 μm.

3.5.2. Rooting Process and Rooting Type of 'HN' Softwood Cuttings

The rooting process of 'HN' cuttings were observed by taking the cuttings prepared at 10 DBF as an example (Figure 5). Initially, the leaflets of 'HN' cuttings were purplish red and curled on 0~10 days after cutting (DAC) (Figure 5A), then turned green on 20~30 DAC (Figure 5B) and unfolded after 40 DAC (Figure 5C). Synchronously, the stem of cuttings gradually turned from purplish red to green and yellow (Figure 5A–D). Generally, the callus tissues were primarily formed at the base incision of cuttings on 10 DAC (Figure 5E) and gradually enlarged at 10~50 DAC (Figure 5F); then, the callus nodules were prominently formed and observed on the surface of callus at 56 DAC (Figure 5G). Finally, the adventitious roots protruded from callus nodules of the cutting base or the stem bark of cutting beyond 60 DAC (Figure 5H,I).

Figure 5. The adventitious root formation of softwood cuttings of the tree peony 'High Noon'. (**A**) The cutting status on 0 DAC. (**B**) Leaves turned green on 28 DAC. (**C**) Leaves unfolding on 56 DAC. (**D**) Adventitious root formation at 62 DAC. (**E**) The callus at the base incision at 10 DAC. (**F**) Callus tissues enlarging at 22 DAC. (**G**) Nodules formed from callus at 56 DAC. (**H**,**I**) ARs out of the callus or bark at 62 DAC. DAC: days after cutting.

We observed that the rooting of the 'HN' cuttings was divided into three types. Most adventitious roots generated from callus tissues and a few were generated from stem bark, while some demonstrated integrated root induction (Figure 6A–C). Through section observation on the anatomical structure, no latent root primordia were found in the stem of softwood cuttings (Figure 6D). In the callus-induced rooting type, the adventitious roots were formed from callus meristem nodules which were differentiated from cortical parenchyma cells (Figure 6E). In the bark-derived rooting type, the cells of vascular cambium layer or phloem parenchyma cells dedifferentiated to root primordia and developed into ARs (Figure 6F).

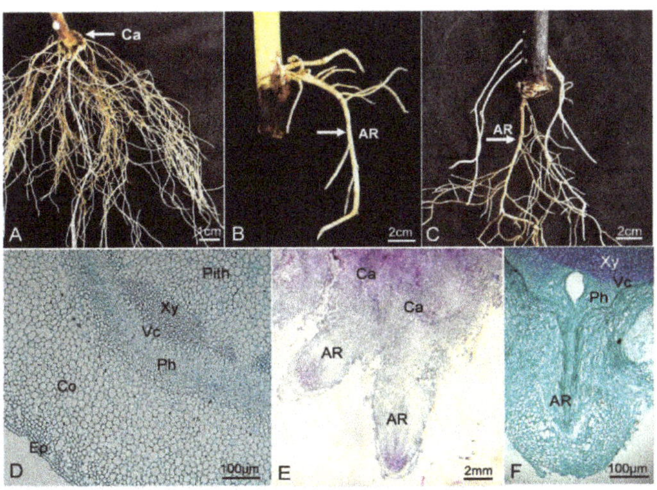

Figure 6. The rooting type of softwood cuttings of tree peony. (**A**) Callus induction; (**B**) bark-derived rooting; (**C**) integrated root induction; (**D**) the status of cutting at 0 DAC; (**E**) the section observation

of AR from the callus; (**F**) the section observation of AR out of the vascular cambium. Abbreviations: Ca, callus tissue; AR, adventitious root; Ep, epidermis; Co, cortex; Ph, phloem; Vc, vascular; Xy, xylem.

4. Discussion

Rooting of cuttings is a prerequisite for the establishment of post-field transfer survival for plants in the soil [45]. Previous reports showed that tree peony cuttings find it difficult to root and that the rooting capacity of this species is significantly different among cultivars [38,46], although 'High noon' softwood cuttings find it relatively easier to root [39]. In this study, nine cultivars demonstrated significant differences in rooting ability, among which 'High noon' and 'Jinghua Qingxue' demonstrated better rooting parameters than other cultivars. The rooting ability of cuttings varies greatly according to the developmental stages and anatomical structure of plants [7,43]. For this reason, the shoots with higher rooting ability should be collected for cuttings which are more conducive to rooting and survival [43]. We found that the highest rooting rates of 'High noon' and 'Jinghua Qingxue' cuttings were obtained frox 28 DAF (50%) and 19 DAF (53.33%) in June when the shoots were in a vigorous growth period (Table 3), which was similar to the rooting result of *Brunfelsia pauciflora* and *Cercis siliquastrum* [24,47]. It was observed that the phloem and xylem parts of shoots were obviously thickened and that a ring of sclerenchym formed outside the phloem of the 'HN' softwood after 55 DAF (Figure 4c,d). It was proposed that when the shoots gradually undergo lignification, root-promoting substances might progressively diminish, while those with inhibitory effects continuously accumulate, which would result in decreased rooting ability [16,23].

Currently, the application of IBA is found to be more effective for promoting AR formation in difficult-to-root cuttings than other plant growth regulators [7,20]. However, the effect of IBA on rooting depends on the genotypes and phenological stages of shoots [5,12,19]. The effective treatments with IBA were 300 mg·L^{-1} for 1 h in *Acer Rubrum* [21] and 4000 mg·L^{-1} for 10 s in *Brunfelsia pauciflora* [24] for stem cuttings. In this study, the highest rooting rate of 'High noon' was obtained using IBA 2000 mg·L^{-1}, followed by 1000 mg·L^{-1} (Figure 1A). The rooting powder induced better performance in terms of root number and root length, followed by IBA 2000 mg·L^{-1} (Figure 1B,C). Therefore, the combination of IBA might need to be optimized to improve rooting for softwood cuttings in tree peony.

Polyamines combined with IBA could affect the rooting performance in stem cuttings, which produced higher rooting rate and better root quality than polyamines alone in woody plants [28,48,49]. Wen et al. [48] found that Put was more effective at promoting in the vitro rooting of three tree peony cultivars than Spm and Spd in combination with IBA, while the optimal treatment of Put varied among the three cultivars. The synergistic effects of IBA with Put, Spd, and Spm could all promote the callus rate of cuttings, but the latter two had no significant effects on the rooting of 'Jinghua Qingxue' (Figure 2A). Moreover, Put 1.0 mM could induce the maximum rooting rate of 25% along with IBA 2000 mg·L^{-1}, and both the root number and root length reached their best values of 6.33 and 4.0 cm, respectively (Figure 2B,C). The role of PAs in the initiation and induction phases of rooting has been previously reported [49,50], while the interaction of PAs and auxin has been hypothesized during the AR formation of stem cuttings [13,28]. Moreover, a lower content of free PAs was observed in difficult-to-root species than in easy-to-root species, indicating the positive role of PAs in rooting [13,29]. In the future, the combination of PAs and IBA treatments should be adjusted to improve rooting rates and reduce rot rates in tree peony softwood cuttings, and the effect of PAs on rooting also need to be analyzed further.

The optimal substrate should have high water holding capacity and sufficient porosity for achieving the right air/water balance [14], while the ideal substrates depend on plant genotypes, the cutting's phenological stage, and the auxin types [32,51,52]. Recently, propagators have tended to use a combination of organic (cocopeat, peat, peat moss, sand) and mineral components (perlite, vermiculite, rockwool) as substrates for soilless cultivation [11,51–55]. Abass et al. [51] found that the rooting rates and root number both performed better in sand/perlite (*v:v* 1:1) and sand/vermiculite (*v:v* 1:1) combined with IBA 3000 $mg \cdot L^{-1}$ for cuttings of *Pedilanthus tithymaloids*. Ghimire et al. [52] reported that a substrate of perlite/vermiculite (*v:v* 1:1) with IBA 2000 $mg \cdot L^{-1}$ was more effective than soil in promoting rooting in cuttings of *Chrysanthemum indicum* L. In this study, IBA 2000 $mg \cdot L^{-1}$ combined with T2, T6, and T4 could significantly increase the rooting rates of 'Jinghua Qingxue' cuttings, reaching 54.17%, 53.33%, and 37.5%, respectively (Figure 3A). Similarly, higher callus rates were obtained from T2, T4, T5, and T6, ranging from 91.67% to 96.67% (Figure 3A). As reported, perlite has higher water holding capacity and pores and lower bulk density than vermiculite [56], so substrates including perlite were more conducive to rooting [57], which was consistent with our results. It was revealed that the rot rate of T2 was higher than T6, but root number and length performed better in T2 than in T6 (Figure 3B,C). Better root formation was attributed to higher water holding capacity and better aeration for cutting substrates [52]. Based on the analysis, the T2 substrate was more suitable for promoting rooting of softwood cutting in the tree peony 'JHQX'. In the future, the physical and chemical properties of substrates should be analyzed before the cutting is planted; meanwhile more combinations of different substrate mixtures and auxin application need to be tested to increase the rooting performance of tree peony softwood cuttings.

AR formation can be divided into three types, namely callus induction, bark-derived rooting, and integrated rooting induction [58,59]. In this study, most cuttings rooted from the callus, but a few were derived from bark, and some were of the integrated rooting type, which was consistent with the results of Chen et al. [39]. As reported by Cheng et al. [60], ARs mainly originated from meristematic nodules of the callus which were formed from the differentiation of cortical parenchyma cells. We found that the callus tissues of cuttings were primarily formed on 10 DAC and gradually enlarged at 10~50 DAC during rooting; ARs then protruded through the callus surface beyond at least 60 DAC (Figure 5). Through the anatomical observation (Figure 6), no root primordia were found in the stem of the tree peony before cutting, and root primordia originated from callus meristem nodules, in accordance with the result of Cheng et al. [60] Additionally, roots were also derived from vascular cambium cells or phloem parenchyma cells, which was similar to the research of Meng et al. [40] and Wen et al. [42]. These findings are consistent with the results of previous studies [37–40], indicating that the rooting of tree peony cuttings mainly belongs to the callus induction type and that tree peony is a difficult-to-root species.

According to the present study, an attempt at acclimation and transplanting for seedlings was made using the rooted cuttings, and a propagation protocol was established for softwood cuttings of tree peony, as illustrated in Figure 7. Firstly, the current shoots after pruning rejuvenation were collected and pruned for cuttings, then sterilized with fungicides (e.g., carbendazim) and inserted into the appropriate substrate (Figure 7A–C). The cuttings were artificially water-sprayed and we applied fungicides and pesticide during the rooting, as described in Section 2.2 (Figure 7D). At about 100 DAC, the cuttings could form root balls and were transplanted into suitable pots (Figure 7E,F). During the rooting cuttings' growth, a water supply and fertilizer input were required regularly. After one year of being transplanted, vigorous roots and newly grown leaves could be seen (Figure 7G). After 30 months of being transplanted, flower buds appeared at the shoot apex (Figure 7H). Consequently, softwood cutting could be used as an effective method to promote AR formation and improve reproductive efficiency for tree peony propagation.

Figure 7. The propagation protocol of softwood cuttings in the tree peony 'High Noon'.

5. Conclusions

The study showed that two cultivars of tree peony, namely 'High noon' and 'Jinghua Qingxue', demonstrated better rooting ability than other cultivars. The highest rooting rates of both cultivars were obtained from shoots at the vigorous developmental stages in June. Moreover, IBA 2000 mg·L^{-1} was proved to significantly promote the rooting of 'High noon' cuttings. Put 1.0 mM combined with IBA 2000 mg·L^{-1} could induce a higher rooting rate and a better rooting index for 'Jinghua Qingxue' cuttings, alongside a recommended substrate of cocopeat/perlite (*v:v* 1:1). The cuttings of the tested cultivars mainly demonstrated callus induction rooting, and most ARs were generated from callus meristem nodules differentiated from cortical parenchyma cells. Based on these results, a protocol propagation was primarily established for softwood cuttings of tree peony.

Author Contributions: Conceptualization, D.L. and F.C.; Methodology, D.L., F.C. and Y.Z.; Formal analysis, D.L.; Funding acquisition, F.C.; Investigation, D.L. and X.T.; Resources, X.T. and F.C.; Data curation, D.L., F.C. and Y.Z.; Writing—original draft preparation, D.L. and F.C.; Writing—review and editing, D.L. and F.C. All authors have read and agreed to the published version of the manuscript.

Funding: This research was funded by the National Key Research and Development Project of China (Grant No. 2020YFD1000503) and the 5·5 Engineering Research & Innovation Team Project of Beijing Forestry University (Grant No. BLRC2023A06).

Data Availability Statement: Data are contained within this article.

Conflicts of Interest: Author Xiwen Tao was employed by the Beijing Guose Peony Technology Co., Ltd. The remaining authors declare that the research was conducted in the absence of any commercial or financial relationships that could be construed as a potential conflict of interest.

References

1. Li, P.; Shen, J.; Wang, Z.; Liu, S.; Liu, Q.; Li, Y.; He, C.; Xiao, P. Genus *Paeonia*: A comprehensive review on traditional uses, phytochemistry, pharmacological activities, clinical application, and toxicology. *J. Ethnopharmacol.* **2021**, *269*, 113708. [CrossRef] [PubMed]
2. Yan, G.; Zhang, L.; Wu, D.; Jiang, S.; Wu, Q.; Dai, M. Paeonol attenuates nonalcoholic steatohepatitis by regulating intestinal flora and AhR/NLRP3/Caspase-1 metabolic pathway. *J. Ethnopharmacol.* **2024**, *329*, 118147. [CrossRef] [PubMed]
3. Cheng, F.; Li, J.; Chen, D.; Zhang, Z. *Chinese Flare Mudan*; China Forestry Publishing House: Beijing, China, 2005; Volume 5, pp. 44–46.

4. John, M.D.; James, L.G. *Cutting Propagation: A Guide to Propagating and Producing Floriculture Crops*; Ball Publishing: Batavia, IL, USA, 2006; pp. 3–4.
5. Monder, M.J.; Pacholczak, A. Rhizogenesis and concentration of carbohydrates in cuttings harvested at different phenological stages of once-blooming rose shrubs and treated with rooting stimulants. *Biol. Agric. Hortic.* **2020**, *36*, 53–70. [CrossRef]
6. Morales-Orellana, R.J.; Winkelmann, T.; Bettin, A.; Rath, T. Stimulation of adventitious root formation by laser wounding in rose cuttings: A matter of energy and pattern. *Front. Plant Sci.* **2022**, *13*, 1009085. [CrossRef]
7. Hejazi, Z.; Safi, S.; Barakzai, A.L.; Karimi, B.K.; Qurashi, F.M.; Tetsumura, T. Cultivar effect on root development in pomegranate hardwood cuttings treated with auxin. *Rhizosphere* **2023**, *25*, 100661. [CrossRef]
8. Li, S. Molecular bases for the regulation of adventitious root generation in plants. *Front. Plant Sci.* **2021**, *12*, 614072. [CrossRef]
9. Guan, L.; Murphy, A.; Peer, W.; Gan, L.; Li, Y.; Cheng, Z. Physiological and molecular regulation of adventitious root formation. *Crit. Rev. Plant Sci.* **2015**, *34*, 506–521. [CrossRef]
10. Johnson, E.P.; Preece, J.E.; Aradhya, M.; Gradzeil, T. Rooting response of *Prunus* wild relative semi-hardwood cuttings to indole-3-butyric acid potassium salt (KIBA). *Sci. Hortic.* **2020**, *263*, 109144. [CrossRef]
11. Pan, T.; Chen, X.; Hao, Y.; Jiang, C.; Wang, S.; Wang, J.; Wei, Q.; Chen, S.; Yu, X.; Cheng, F.; et al. Optimization of factors affecting the rooting of pine wilt disease resistant Masson pine (*Pinus massoniana*) stem cuttings. *PLoS ONE* **2021**, *16*, e0251937. [CrossRef]
12. Kumar, A.; Nagar, P.K.; Palni, L.M.S. The role of indole-3-butyric acid and polyamines on in vitro rooting of microshoots of *Rosa damascena* Mill. and Gladiolus hybridus Hort. *S. Afr. J. Bot.* **2024**, *172*, 1–7. [CrossRef]
13. Tsafouros, A.; Frantzeskaki, A.; Assimakopoulo, A.; Roussos, P.A. Spatial and temporal changes of mineral nutrients and carbohydrates in cuttings of four stone fruit rootstocks and their contribution to rooting potential. *Sci. Hortic.* **2019**, *253*, 227–240. [CrossRef]
14. Hartmann, H.P.; Kester, D.E.; Davies, J.T. *Plant Propagation: Principle and Practices*; Prentice Hall: Hoboken, NJ, USA, 1990.
15. Kostas, S.; Hatzilazarou, S.; Pipinis, E.; Vasileiadis, A.; Magklaras, P.; Smyrnioudis, I.; Vasilakis, T.; Chazakis, M.; Anastasiadi, V.; Ziogou, F.T. Propagation of *Pistacia lentiscus* var. Chia genotypes and determination of their ornamental traits combined with a genetic analysis using ISSR markers. *Agronomy* **2021**, *11*, 205. [CrossRef]
16. Izhaki, A.; Yitzhak, Y.; Blau, T.; David, I.; Rotbaum, A.; Riov, J.; Zilkah, S. Rooting of cuttings of selected *Diospyros virginiana* clonal rootstocks and bud growth in rooted cuttings. *Sci. Hortic.* **2018**, *232*, 13–21. [CrossRef]
17. Sabatino, L.; D'Anna, F.; Iapichino, G. Improved propagation and growing techniques for Oleander nursery production. *Horticulturae* **2019**, *5*, 55. [CrossRef]
18. Yang, X.; Li, S.; Liu, F.; Rosenqvist, E. Does constant or changing light give the best rooting of hibiscus cuttings of two sizes? *Sci. Hortic.* **2023**, *309*, 111675. [CrossRef]
19. Nasri, F.; Fadakar, A.; Saba, M.K.; Yousefi, B. Study of indole butyric acid (IBA) effects on cutting rooting improving some of wild genotypes of damask roses (*Rosa damascena* Mill.). *J. Agric. Sci.* **2015**, *60*, 263–275. [CrossRef]
20. Azad, M.S.; Alam, M.J.; Mollick, A.S.; Khan, M.N.I. Rooting of cuttings of the wild Indian almond tree (*Sterculia foetida*) enhanced by the application of indole-3-butyric acid (IBA) under leafy and non-leafy conditions. *Rhizosphere* **2018**, *5*, 8–15. [CrossRef]
21. Zhu, W.; Zhang, M.; Li, J.; Ge, W. Key regulatory pathways, microRNAs, and target genes participate in adventitious root formation of *Acer rubrum* L. *Sci. Rep.* **2022**, *12*, 12057. [CrossRef]
22. Loconsole, D.; Sdao, A.E.; Cristiano, G.; Lucia, B.D. Different responses to adventitious rhizogenesis under indole-3-butyric acid and seaweed extracts in ornamental's cuttings: First Results in *Photinia x fraseri* 'Red Robin'. *Agriculture* **2023**, *13*, 513. [CrossRef]
23. Song, X.; Huang, R.; Liu, H.; Zhang, J.; Chang, Y.; Pei, D. Transcriptome profiling of indole-3-butyric acid–induced adventitious root formation in softwood cuttings of walnut. *Hortic. Plant J.* **2024**, *10*, 1336–1348. [CrossRef]
24. Sousa, C.E.D.; Oliveria, F.L.R.; Sant'Anna-Santos, B.F.; Zuffellato-Ribas, K.C. Physiological and anatomical aspects of the rooting of *Brunfelsia pauciflora* cuttings. *Sci. Hortic.* **2023**, *307*, 111491. [CrossRef]
25. Frick, E.M.; Strader, L.C. Roles for IBA-derived auxin in plant development. *J. Exp. Bot.* **2018**, *69*, 169–177. [CrossRef]
26. Li, S.; Huang, P.; Ding, G.; Zhou, L.; Tang, P.; Sun, M.; Zheng, Y.; Lin, S. Optimization of hormone combinations for root growth and bud germination in Chinese fir (*Cunninghamia lanceolata*) clone leaf cuttings. *Sci. Rep.* **2017**, *7*, 5046. [CrossRef]
27. Zhou, X.; Li, R.; Shen, H.; Yang, L. Effect of Exogenous Plant Growth Regulators and Rejuvenation Measures on the Endogenous Hormone and Enzyme Activity Responses of *Acer mono* Maxim in Cuttage Rooting. *Int. J. Mol. Sci.* **2023**, *24*, 11883. [CrossRef]
28. Cristofori, V.; Rouphael, Y.; Rugini, E. Collection time, cutting age, IBA and putrescine effects on root formation in *Corylus avellana* L. cuttings. *Sci. Hortic.* **2010**, *124*, 189–194. [CrossRef]
29. Denaxa, N.K.; Roussos, P.A.; Vemmos, S.N. The possible role of polyamines to the recalcitrance of "Kalamata" olive leafy cuttings to root. *J. Plant Growth Regul.* **2014**, *33*, 579–589. [CrossRef]
30. Tsafouros, A.; Roussos, P.A. The possible bottleneck effect of polyamines' catabolic enzymes in efficient adventitious rooting of two stone fruit rootstocks. *J. Plant Physiol.* **2020**, *244*, 152999. [CrossRef]

31. Niu, L.; Tang, Y.; Zhu, B.; Huang, Z.; Wang, D.; Chen, Q.; Yu, J. Nitric oxide promotes adventitious root formation in cucumbe under cadmium stress through improving antioxidant system, regulating glycolysis pathway and polyamine homeostasis. *Fron Plant Sci.* **2023**, *14*, 1126606. [CrossRef]
32. Campbell, S.M.; Anderson, S.L.; Brym, Z.T.; Pearson, B.J. Evaluation of substrate composition and exogenous hormone applicatio on vegetative propagule rooting success of essential oil hemp (*Cannabis sativa* L.). *PLoS ONE* **2021**, *16*, e0249160. [CrossRef]
33. Nemati, R.; Fortin, J.P.; Craig, J.; Donald, S. Growing Mediums for Medical Cannabis Production in North America. *Agronom* **2021**, *11*, 1366. [CrossRef]
34. Weingarten, M.; Mattson, N.; Grab, H. Evaluating Propagation Techniques for *Cannabis sativa* L. Cultivation: A Comparativ Analysis of Soilless Methods and Aeroponic Parameters. *Plants* **2024**, *13*, 1256. [CrossRef] [PubMed]
35. Gopinath, P.; Vethamoni, P.I.; Gomathi, M. Aeroponics soilless cultivation system for vegetable crops. *Chem. Sci. Rev. Lett.* **2017**, *6* 838–849.
36. Eldridge, B.M.; Manzoni, L.R.; Graham, C.A.; Rodgers, B.; Farmer, J.R.; Dodd, A.N. Getting to the roots of aeroponic indoo farming. *New Phytol.* **2020**, *228*, 1183–1192. [CrossRef]
37. Blaney, L.T. Tree peonies rooted from leaf-bud cuttings. *Or. Ornam. Nurs. Dig.* **1958**, *2*, 1+2+4.
38. Zeng, D.; Yin, W.; Wang, Y.; Zhao, X.; Wang, H. Propagation with Etiolated Softwood Cuttings of Five Dwarf Cultivars of Chines Tree Peony. *Acta Hortic. Sin.* **2005**, *32*, 725.
39. Chen, C. Study on the Cutting Rooting and Stem Anatomical Structure Development of Tree Peony. Master's Thesis, Beijin Forestry University, Beijing, China, 2019.
40. Meng, X.; Wang, Z.; He, S.; Shi, L.; Song, Y.; Lou, X.; He, D. Endogenous hormone levels and activities of IAA-modifying enzyme during adventitious rooting of tree peony cuttings and grafted scions. *Hortic. Environ. Biotechnol.* **2019**, *60*, 187–197. [CrossRef]
41. Xu, L.; Cheng, F.; Zhong, Y. Histological and cytological study on meristematic nodule induction and shoot organogenesis i *Paeonia ostii* 'Feng Dan'. *Plant Cell Tiss. Org.* **2022**, *149*, 609–620. [CrossRef]
42. Wen, S.; Miao, D.; Cui, H.; Li, S.; Gu, Y.; Jia, R.; Leng, Y. Physiology and transcriptomic analysis of endogenous hormone regulating in vitro adventitious root formation in tree peony. *Sci. Hortic.* **2023**, *318*, 112122. [CrossRef]
43. Monder, M.J.; Niedzielski, M.; Woliński, K. The pivotal role of phenological stages enhanced by plant origin preparations in th process of rhizogenesis of *Rosa* 'Hurdal'stem cuttings. *Agriculture* **2022**, *12*, 158. [CrossRef]
44. Tong, N.; Shu, Q.; Wang, B.; Peng, L.; Liu, Z. Histology, physiology, and transcriptomic and metabolomic profiling reveal th developmental dynamics of annual shoots in tree peonies (*Paeonia suffruticosa* Andr.). *Hortic. Res.* **2023**, *10*, uhad152. [CrossRef
45. Uwe, D.; Alexander, H.; Manuel, P.P.J.; Yvonne, K.; Manuel, A.; Fahimeh, S.; Siegfried, Z.; Philipp, F.; Hajirezaei, M.R. Molecula and physiological control of adventitious rooting in cuttings: Phytohormone action meets resource allocation. *Ann. Bot.* **2019**, *12* 929–949.
46. Shi, Q.; Wang, Y.; Zhou, L.; Huang, G. Influence of Rooting Powder and Culture Substrata on Rooting of Different Traditiona Cultivars of *Paeonia suffruticosa*. *North. Hortic.* **2012**, *1*, 91–96.
47. Karam, N.S.; Gebre, G.H. Rooting of *Cercis siliquastrum* cuttings influenced by cutting position on the branch and indole-butyri acid. *J. Hortic. Sci. Biotechnol.* **2004**, *79*, 792–796. [CrossRef]
48. Wen, S.; Cheng, F.; Zhong, Y.; Wang, X.; Li, L.; Zhang, Y.; Qiu, J. Efficient protocols for the micropropagation of tree peony (*Paeoni suffruticosa* 'Jin Pao Hong', *P. suffruticosa* 'Wu Long Peng Sheng', and *P.* × *lemoinei* 'High Noon') and application of arbuscula mycorrhizal fungi to improve plantlet establishment. *Sci. Hortic.* **2016**, *201*, 10–17. [CrossRef]
49. Denaxa, N.K.; Vemmos, S.N.; Roussos, P.A. Shoot girdling improves rooting performance of kalamata olive cuttings by upregulat ing carbohydrates, polyamines and phenolic compounds. *Agriculture* **2021**, *11*, 71. [CrossRef]
50. Matam, P.; Parvatam, G. Putrescine and polyamine inhibitors in culture medium alter in vitro rooting response of *Decalepi hamiltonii* Wight&Arn. *Plant Cell Tiss. Org.* **2017**, *128*, 273–282.
51. Abass, M.M.; Thabet, R.S.; Lasheen, F.F.; Abdelhamid, A.N.; Hassan, K.M.; Saudy, H.S.; Boghdady, M.S. Modulating th rhizosphere medium and indole−3−butyric acid supply influence rooting, nutrients and biochemical constituents and histologica features of *Pedilanthus tithymaloids*. *J. Soil Sci. Plant Nut.* **2024**, *24*, 6880–6892. [CrossRef]
52. Ghimire, B.K.; Kim, S.H.; Yu, C.Y.; Chung, I.M. Biochemical and physiological changes during early adventitious root formatio in *Chrysanthemum indicum Linné* cuttings. *Plants* **2022**, *11*, 1440. [CrossRef]
53. Malviya, N.K.; Chaurasiya, R.; Maji, S. Use of cocopeat for soilless cultivation of tomato. *South Asian J. Exp. Biol.* **2020**, *10*, 169–17 [CrossRef]
54. Baronti, S.; Montagnoli, A.; Beatrice, P.; Danieli, A.; Maienza, A.; Vaccari, F.P.; Casini, D.; Di Gennaro, S.F. Above- and below ground morpho-physiological traits indicate that biochar is a potential peat substitute for grapevine cuttings nursery productio *Sci. Rep.* **2024**, *14*, 17185. [CrossRef]
55. Erdal, İ.; Hakan, A. Comparison of the perlite, leonardite, vermicompost and peat moss and their combinations with cocopeat a tomato growing media. *J. Soil Sci. Plant Nutr.* **2025**. [CrossRef]

56. He, M.; Lv, L.; Li, H.; Meng, W.; Zhao, N. Analysis on soil seed bank diversity characteristics and its relation with soil physical and chemical properties after substrate addition. *PLoS ONE* **2016**, *11*, e0147439. [CrossRef] [PubMed]
57. Vijayaraghavan, K.; Franklin, D.R. Design and development of green roof substrate to improve runoff water quality: Plant growth experiments and adsorption. *Water Res.* **2014**, *63*, 94–101. [CrossRef] [PubMed]
58. Xu, L. De novo root regeneration from leaf explants: Wounding, auxin, and cell fate transition. *Curr. Opin. Plant Biol.* **2018**, *41*, 39–45. [CrossRef]
59. Guan, L.; Li, Y.; Huang, K.; Cheng, Z. Auxin regulation and *MdPIN* expression during adventitious root initiation in apple cuttings. *Hortic. Res.* **2020**, *7*, 143. [CrossRef]
60. Cheng, F.Y.; Wang, Y.P. The propagation method of peony by its twig and the cytohistological observation on the root regeneration. *Acta Hortic. Sin.* **1993**, *20*, 176–180.

horticulturae

MDPI

Article

Seed Dormancy and Germination Characteristics of *Scutellaria indica* L. var. *coccinea* S.T.Kim & S.T.Lee., an Endemic Species Found on Jeju Island, South Korea

Jae Hui Kim [1], Hak Cheol Kwon [2] and Seung Youn Lee [1,3,*]

[1] Department of Horticulture and Breeding, Graduate School of Gyeongkuk National University, Andong 36792, Republic of Korea; wogml0157@gmail.com
[2] Natural Product Informatics Research Center, Korea Institute of Science and Technology, Gangneung 25451, Republic of Korea; hkwon@kist.re.kr
[3] Major of Smart Horticultural Science, Gyeongkuk National University, Andong 36729, Republic of Korea
* Correspondence: mrbig99@gknu.ac.kr; Tel.: +82-54-820-5472

Abstract

Globally, biodiversity is declining, meaning that many endemic plants are under threat; therefore, it is essential to develop conservation strategies. *Scutellaria indica* var. *coccinea* has great potential as an ornamental ground cover plant, but it is a plant that requires ex situ conservation. This study was conducted in order to investigate the seed germination characteristics and classify the dormancy type of *S. indica* var. *coccinea* seeds, with the aim of developing mass propagation protocols for ex situ conservation and preservation of their genetic diversity. Fresh and mature seeds of *S. indica* var. *coccinea* are in a dormant state, which is released by low temperatures during winter, resulting in radicle and seedling emergence the following spring. At the time of dispersal, the seeds had fully developed embryos, and the seed coat was permeable. When the seeds were incubated under four different temperature regimes (4, 15/6, 20/10, or 25/15 °C), they showed a low germination percentage ($\leq$20%), indicating that a substantial proportion of the seeds remained in a dormant state. In the cold stratification experiment (0, 4, 8, or 12 weeks at 4 °C), germination increased, and the time required for germination shortened as the duration of cold treatment lengthened. This suggests that low temperatures are the primary environmental signal that induces germination. In the gibberellic acid (GA_3) treatment (GA_3 0, 10, 100, or 1000 mg·L^{-1}), relatively high concentrations (100 and 1000 mg·L^{-1}) were effective in promoting germination. The highest germination was obtained in GA_3 1000 mg·L^{-1} with 100.0%, which is about 7 times higher than the control (13.6%). Therefore, seeds of *S. indica* var. *coccinea* were classified as having non-deep physiological dormancy (PD). Additionally, because the minimum germinable temperature range of the seeds was extended to lower temperatures, the seeds were classified as having type 2 non-deep PD.

Keywords: cold stratification; dormancy break; endemic plant; GA_3; phenology; *Scutellaria*; seed germination

1. Introduction

Biodiversity refers to the variety of life, including the number of species, genetic variation, and ecological and evolutionary processes, and it plays a crucial role in maintaining the stability and balance of ecosystems [1–3]. Plants play a vital role in the response to and mitigation of climate change by absorbing carbon and reducing greenhouse gases in the atmosphere [4,5]. Therefore, they must be preserved not only for environmental protection but also for both the survival and future of humanity. However, plant biodiversity continues to decline due to human activities (overharvesting and urbanization), habitat changes, invasion of alien species, and climate change [6–10].

South Korea, located in East Asia, has relatively abundant flora and diverse vegetation due to its mountainous terrain (≥62%), climate, and geographical features [11,12]. However, the forest ecosystems of South Korea are threatened by rapid industrialization, overharvesting, and pollution [13]. In order to protect them, the Ministry of Environment, the Ministry of Agriculture, Food, and Rural Affairs, and the Korea Forest Service are designating protected areas through conservation agencies, implementing in situ and ex situ restoration programs for species, and strengthening legal protection and regulations [12,14].

Endemic plants, resulting from genetic diversity, have adapted to the specific environmental conditions of a localized area and are naturally found only in limited areas [15–19]. These plants contribute to the maintenance of biodiversity through their unique genetic traits, and typically exist in small populations, making it necessary to manage and protect them to prevent extinction [5,20]. In South Korea, various ex situ conservation institutes, such as the Korea National Arboretum (KNA), National Institute of Biological Resources (NIBR), and botanical gardens, are operating to protect threatened plant species [21,22]. Among these, botanical gardens are one of the simplest ex situ conservation strategies to secure the safety of threatened plants in their natural habitats. They conduct research on plant protection and conservation and develop propagation methods [10,20,23]. Objective plant propagation techniques based on scientific results are required for botanical gardens to effectively conserve rare or endemic plants ex situ.

Scutellaria is the second largest genus in the Lamiaceae family, consisting of annual or perennial herbaceous plants [24,25], and is mainly distributed in East Asia, including Korea, Japan, Taiwan, and China [26–29]. The genus *Scutellaria* contains flavonoids, phenolic compounds, and amino acids and is known for its high medicinal value, including anticancer, anti-inflammatory, and antihypertensive effects [30–32]. Additionally, *S. ocmulgee* and *S. suffrutescens*, which are native to North America, have a long flowering period of 2–5 months, and they have been reported to have potential ornamental value because they produce many flowers [29]. *Scutellaria*, which is distributed throughout Korea, has 10 species and eight different varieties, including *S. dependens*, *S. insignls*, *S. indica* var. *tsusimensis*, and *S. indica* var. *coccinea*, etc. A new variant of Korean *Scutellaria*, *S. indica* var. *coccinea*, is an endemic plant that was found in a small volcanic crater called 'Sangumburi' on Jeju Island [18,25,26]. Jeju Island Sangumburi has been designated as a natural monument and national heritage site, and its surrounding plants are also protected [33]. This perennial herbaceous plant is known for its medicinal properties, including effects on active blood, detoxification, and bruising [34]. It grows to approximately 20–30 cm in height and it blooms from May to July with deep pink flowers [26,35,36], making it potentially suitable for ornamental use. However, because it exists in very small populations, it is particularly vulnerable to climate and deforestation; therefore, it is important to develop conservation techniques to protect it. The Korea Forest Service has enforced the Forest Protection Act to protect plants from their natural habitats, but no research has been conducted on their ex situ conservation [37].

The genetic diversity of plants can be conserved through methods such as seed and vegetative propagation as well as plant tissue culture [38]. Seed propagation is easy to perform and maintains genetic diversity during storage [39]. Understanding the early life history of a plant is important for seed propagation. Seed germination and seedling establishment are the first steps in a plant's life cycle [40]. Dormancy refers to the state in which seeds do not germinate for a certain period, even when environmental conditions such as temperature and light are favorable for germination [41,42]. Although the seeds in this state are viable and intact, they cannot germinate for several days or months. Plants belonging to the same genus or family may exhibit similar types of seed dormancy; however, the germination conditions required to reach the plant may be different [40,43]. Previous research on seed dormancy and germination characteristics of several species belonging to the Lamiaceae family has been reported. Baskin and Baskin [40] reported that seeds of plants belonging to the genera *Scutellaria* and *Salvia* have physiological dormancy (PD). In *S. parvula*, which is native to the eastern United States, the seeds are in a dormant state at the time of detachment from the parent plant, and dormancy is broken through after-ripening [44]. Seeds of *S. rubropunctata* that remained ungerminated at 30 °C germinated soon after the temperature was changed to 15 °C, indicating a state of enforced dormancy [45]. The seeds of *Salvia aegyptica*, widely distributed in temperate and subtropical climates, germinate across a broad temperature range (10–40 °C), with an optimal germination temperature reported to be 30 °C [46]. In the case of *Teucrium santae*, native to Northwest Algeria, the optimal germination temperature was found to be between 15 and 25 °C, with the highest germination observed at 20 °C [47]. Additionally, *Hyptis suaveolens*, native to the semi-arid region of Northeastern Brazil, exhibited the highest germination at temperatures between 25 and 30 °C [48]. Since germination traits can vary depending on the environmental characteristics of the plant, customized propagation studies are therefore necessary for each species. However, no studies have been conducted to date on the dormancy and germination characteristics of *S. indica* var. *cocinea* seeds native to Jeju Island, so it is important to investigate the germination conditions required for this plant. This study aimed to understand the life cycle and germination characteristics of *S. indica* var. *coccinea* seeds and, if dormancy is present, classify the type of dormancy to develop the techniques necessary for propagation.

The following experiments were conducted to understand the life cycle of the target species and investigate seed germination characteristics: (1) phenology of germination and seedling emergence under natural conditions; (2) temperature and light requirements for germination; (3) effects of different periods of cold stratification on germination; and (4) effects of gibberellic acid on germination in the laboratory. The results of this study can be used to develop a seed-based mass propagation technique for *S. indica* var. *coccinea* and provide fundamental information for ex situ conservation.

2. Materials and Methods

2.1. Plant Materials

The mature seeds of *S. indica* var. *coccinea* with dark brown seed coats were collected from cultivated plants grown from wild-collected seed sources on 21 July 2021, cultivated plants present in the Hantaek Botanical Garden in Gyeonggi-do, Korea, a conservation agency designated by the Korea Forest Service and the Ministry of Environment for ex situ conservation. The collected seeds were dried for 2 weeks at room temperature (23 ± 2 °C). The dried seeds were then sealed in a sterile plastic bottle with silica gel and stored in cold storage (1 °C) until they were required for the experiments.

2.2. Basic Characteristics of Seed

Randomly selected seeds were measured for characteristics including size, weight, and internal and external shapes. Twenty seeds were measured using Vernier calipers (Digital caliper 150 mm, Kangalu, Shenzhen, Guangdong, China). The weight of 100 seeds was measured in triplicate using an electronic balance (PGA214, OHAUS Corporation, Parsippany, NJ, USA). The external morphology of the seeds was observed after cutting the seeds in half with a razor blade, and the internal morphology was observed using a USB electron microscope (AM3111 Dino-Lite Premier; ANMO Electronics Co., Hsinchu, Taiwan).

2.3. *Phenology Experiments*

In order to assess the seed germination patterns and life cycle under natural conditions, two experiments (seed germination and seedling emergence) were conducted in outdoor phenology experiment plots of Gyeongkuk National University, Andong, Korea (36°32′40.41888″ N, 128°48′2.67516″ E). For the seed germination experiment, a 9 × 8 cm (diameter × height) plastic pot was filled with horticultural soil (Baroker; cocopeat: 68%, peat moss: 14.73%, zeolite: 4%, perlite: 7%, vermiculite: 6%, fertilizer: 0.201%, wetting agent: 0.064%, and pH regulator: 0.005%), followed by placing 20 seeds in a mesh bag containing sand, sealing the bag, and burying it in the pot at a depth of approximately 3 cm. The pot was buried in the ground to ensure that its height was level with the ground surface. For the seedling emergence experiment, a 9 × 8 cm plastic pot was filled with 2/3 horticultural soil, followed by 1 cm of sand, 20 seeds, and 1 cm of horticultural soil. The pot was buried in the ground to ensure that its height was level with the ground surface. The phenology plot was set up to simulate the natural environment in which the seeds grew, with a 30% shade canopy installed (height of 120 cm) during the summer and a thermal nonwoven fabric covering it during the winter. Considering the timing of seed dispersal, this experiment started on 16 July 2023. This experiment lasted for approximately one year and was checked every two weeks. For the germination experiment, the bags were taken out to check for individuals with emerged radicles, while for the seedling emergence experiment, only the number of seedlings that had emerged above the substrate surface was recorded. The reason why distinct evaluation metrics were applied in the two experiments (seed germination and seedling emergence) is as follows: (1) In the case of germination experiment, since it was only necessary to confirm whether the radicle broke through the seed coat, the seeds were tied in the mesh bag and germination was observed, and (2) in the case of seedling emergence, a slightly different method was applied for observation because tying the seeds in the mesh bag would have hindered the growth and emergence of cotyledons and hypocotyls after germination. A data logger (WatchDog Model 540; Spectrum Technologies, Inc., Plainfield, IL, USA) was installed below the soil (2–3 cm) in the phenology plot and set to measure the soil temperatures every 30 min.

2.4. *Water Absorption Test*

To determine the presence of physical dormancy caused by impermeability of the seed coat, a water absorption experiment was conducted. Distilled water was added to a 50 mL glass beaker to fully submerge the seeds, and this experiment was conducted with 20 seeds in three replicates at room temperature (23 ± 2 °C). Seed weight was measured before water absorption and after 3, 6, 9, 12, 24, 48, and 72 h of water absorption, after removing the seeds from the beaker and thoroughly drying the surface. The water absorption percentages were calculated using the following formula:

$$W_s\ (\%) = [(W_i - W_d)/W_d] \times 100$$

where W_s is the mass of the increased seeds, W_i is the seed mass after a given imbibition interval, and W_d is the initial seed mass at 0 h.

2.5. *Laboratory Experiments*

2.5.1. General Procedure

Seeds used in this experiment were rinsed alternately with distilled water and 70% ethanol for 10 s. They were then placed in 15 mL tubes containing 2000 mg·L^{-1} Benomyl

(Farm Hannong, Seoul, Republic of Korea), a fungicide, solution, and stirred on a shaker for 24 h. The seeds were then rinsed twice with distilled water for 10 s. Two layers of filter paper were placed in Petri dishes (90 × 15 mm), and distilled water was added to keep them moist before the seeds were placed. The experiments were conducted with 20 seeds in four replicates. The photosynthetic photon flux density (PPFD) of the Cold-Lab Chamber (4 °C) and Multi-Room Chamber (15/6, 20/10, and 25/15 °C) averaged approximately 1 and 34 $\mu mol \cdot m^{-2} \cdot s^{-1}$. The Petri dishes were wrapped in Parafilm to prevent water evaporation during this experiment. The Petri dishes were continuously supplied with distilled water when needed to prevent them from drying during this experiment. Germination was defined as when the radicle emerged through the seed coat and protruded by more than 1 mm. During the germination assessment, germinated seeds were immediately removed from the Petri dishes, and dead seeds were discarded and excluded from the final germination calculation. When the seeds split in half at the end of this experiment, seeds with white and firm embryos were considered viable and included in the total number of seeds, and the overall percentage of viable seeds was 80–90%. Final germination percentage was calculated using the following formula:

$$\text{Final germination (\%)} = (n/N) \times 100$$

where n is the number of germinated seeds at the end of the incubation period, and N is the total number of viable seeds at the end of this experiment.

2.5.2. Effect of Light on Germination

An experiment was conducted to assess the germination characteristics of the seeds under different light regimes. The light conditions were incubated in light/dark (12 h photoperiod) at 25/15 °C, and the dark conditions were maintained by wrapping the Petri dishes with two layers of aluminum foil to block light and then incubating them in the same environment as the light conditions. Seed germination was checked weekly for 12 weeks, and under dark conditions, light was blocked in the laboratory during the assessment and monitored using a green LED light source.

2.5.3. Effect of Temperature on Germination

To assess the seed germination characteristics under different incubation temperatures, this experiment was conducted under constant temperature (4 °C) and fluctuating temperature (15/6, 20/10, or 25/15 °C) conditions. These four temperature conditions were used to simulate four seasons in Korea, where 4 °C represents winter, 15/6 °C represents early spring and late autumn, 20/10 °C represents late spring and early autumn, and 25/15 °C represents summer. To maintain constant temperature conditions, a Cold-Lab Chamber (HB-603CM, Hanbaek-Scientific Co., Bucheon-si, Republic of Korea) set to 4 °C and a Multi-Room Chamber (HB-302S-4, Hanbeak-Scientific Co., Bucheon-si, Republic of Korea) set to 15/6, 20/10, or 25/15 °C, were used. Seed germination was checked weekly for 12 weeks, and the criteria and methods for the assessment were the same as those described in Section 2.5.1.

2.5.4. Effect of Cold Stratification on Germination

An experiment was conducted to assess germination characteristics under the cold stratification (CS) treatment. The CS experiments were conducted twice. In the first experiment (Experiment 1), the seeds were treated in a Cold-Lab Chamber (4 °C) for 0, 4, or 8 weeks, respectively. In the second experiment (Experiment 2), the seeds were treated at 4 °C for 0, 4, 8, or 12 weeks, respectively, to investigate the increase in germination rate with longer cold stratification periods. During the CS period, the seeds were placed in

Petri dishes lined with two layers of filter paper, distilled water was added to keep the filter paper constantly moist, and the Petri dishes were sealed with Parafilm to prevent moisture loss during the experiment. After each CS period, the seeds were incubated in a Multi-Room Chamber (25/15 °C). Both experiments were completed when no further changes for more than 3 weeks in germination occurred. Seed germination was checked weekly during this experiment, and the criteria and methods for assessment were the same as those described in Section 2.5.1. At the end of the germination period, the mean germination time (MGT) [49–51], time to reach 50% germination (T_{50}) [52], and germination index (GI) [53] were calculated according to the following formula:

$$\text{MGT (days)} = \sum(n * t)/\sum n$$

where n is the number of newly germinated seeds on survey day t, and t is the time from the beginning of the germination test in hours or days.

$$T_{50}\ \text{(days)} = t_i + [(N/2 - n_i)(t_j - t_i)]/(n_j - n_i)$$

where N is the final number of germination, and n_i and n_j are the cumulative number of seeds germinated by adjacent counts at times t_i and t_j, respectively, when $n_j < N/2 < n_i$.

$$\text{GI} = \sum n_i/t_i$$

where n_i is the number of germinated seeds on survey day t_i.

2.5.5. Effect of GA_3 Treatment on Germination

To assess the seed germination characteristics following GA_3 treatment, the seeds were immersed in GA_3 solution of 10, 100, or 1000 mg·L^{-1} for 24 h, respectively. The control, which was not treated with GA_3, was immersed in distilled water for 24 h. After treatment, the seeds were sterilized using the method described in Section 2.5.1 and incubated in a Multi-Room Chamber at 25/15 °C. Seed germination was checked weekly for 12 weeks, and the criteria and methods for the assessment were the same as those described in Section 2.5.1.

2.6. Statistical Analysis

Data were analyzed using the Statistical Analysis System (SAS 9.4, SAS Institute Inc., Cary, NC, USA). The significance among treatments was tested using analysis of variance (ANOVA), and when significant differences were confirmed, means were compared using Duncan's multiple-range tests at $p < 0.05$ as a post hoc analysis. Graphs were plotted using SigmaPlot 12.5 (SPSS Inc., Chicago, IL, USA).

3. Results

3.1. Basic Characteristics of Seed

The fruits of *S. indica* var. *coccinea* have a schizocarp consisting of four mericarps. Each seed was dark brown and had papillate protuberances on its surface (Figure 1B,C). The seed length and width were approximately 1.27 ± 0.03 mm and 0.82 ± 0.02 mm, respectively, and the average weight of 100 seeds was 51.17 ± 0.54 mg. The embryo was foliate, bent, and fully developed (Table 1 and Figure 1D).

Figure 1. Photos of a native habitat of *Scutellaria indica* L. var. *coccinea* S.T.Kim & S.T.Lee (**A**), a capsule containing four mericarps (i.e., seeds) (**B**), external morphology of a seed (**C**), internal morphology of a seed with a fully developed embryo (**D**), germination stages from the emergence of the radicle to the seedlings (**E**), propagules after the transplanted seedlings in a greenhouse (**F**), flowering from the propagules in a greenhouse (**G**), flowering and capsules (**H**), and seed maturation (**I**) in the field. Scale bars indicate 1 mm.

Table 1. Seed characteristics of *Scutellaria indica* L. var. *coccinea* S.T.Kim & S.T.Lee.

Scientific Name	Collection Location	Collection Date	Length (mm)	Width (mm)	100 Seed Weight (mg)
Scutellaria indica L. var. *coccinea* S.T.Kim & S.T.Lee	Hantaek Botanical Garden (37°09′ N, 127°41′ E)	26 June 2021	1.27 ± 0.03 [1]	0.82 ± 0.02	51.17 ± 0.54 [2]

[1] Values of the length and width indicate mean ± standard error (n = 20). [2] Values indicate mean ± standard error (n = 3).

3.2. Phenology Experiment

In the field environment, the first germination was observed on 30 July 2023, two weeks after burying (8.1%). At this time, the average daily soil temperature is 26.5 °C (Figure 2A,B). No changes were observed from August 2023 to early February of the following year; however, on 3 March 2024, germination began to increase, and by 7 July 2024, the final germination reached 100.0%. The seedlings emerged seven weeks after the first germination on 17 September 2023, with a 4.2% emergence. Similarly, no changes were observed from late September 2023 to mid-March of the following year; however, on 31 March 2024, the seedling emergence rate began to increase, reaching a final rate of 73.7% on 7 July 2024 (Figure 2B).

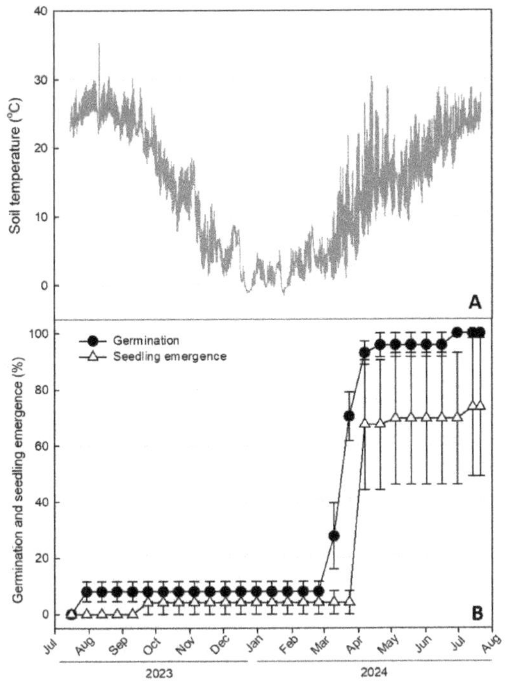

Figure 2. Phenology experiments of *Scutellaria indica* L. var. *coccinea* S.T.Kim & S.T.Lee. Mean daily soil temperature (**A**), cumulative germination, and seedling emergence (**B**). The data was recorded from 16 July 2023 to 21 July 2024. Error bars indicate mean ± standard error of four replicates of 20 seeds.

3.3. Water Absorption Test

The mass of the seed increased by 28.7% after 6 h when compared with the initial weight. The seed absorption curve showed a typical pattern of initial mass increase, followed by a plateau phase and a subsequent increase (Figure 3).

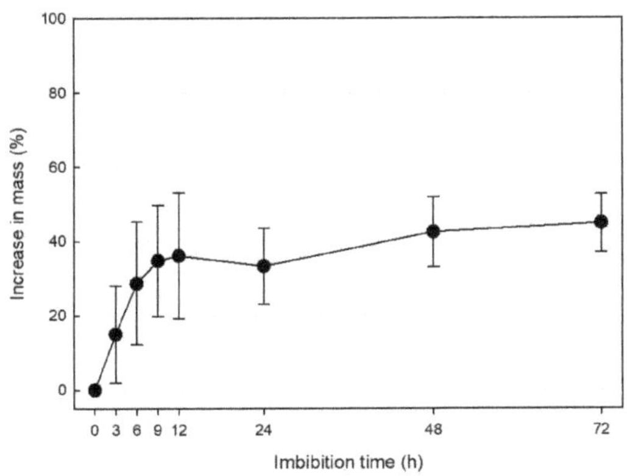

Figure 3. Water uptake by the intact seeds of *Scutellaria indica* L. var. *coccinea* S.T.Kim & S.T.Lee as represented by an increase in mass. Seeds were incubated at room temperature (23 ± 2 °C) in beakers containing distilled water for 72 h. Error bars indicate mean ± standard error of three replicates of 20 seeds.

3.4. Laboratory Experiments

3.4.1. Effect of Light on Germination

The results showed that the first germination was observed after 2 weeks of incubation under light conditions, and after 4 weeks, germination was 8.9% (Figure 4). The final germination rate was 23.3% after 12 weeks of incubation, and a significant number of seeds remained dormant without germination. In the dark, the seeds germinated after 4 weeks of incubation, and the final germination rate was 17.8%. The ANOVA results showed no significant differences between treatments under light conditions.

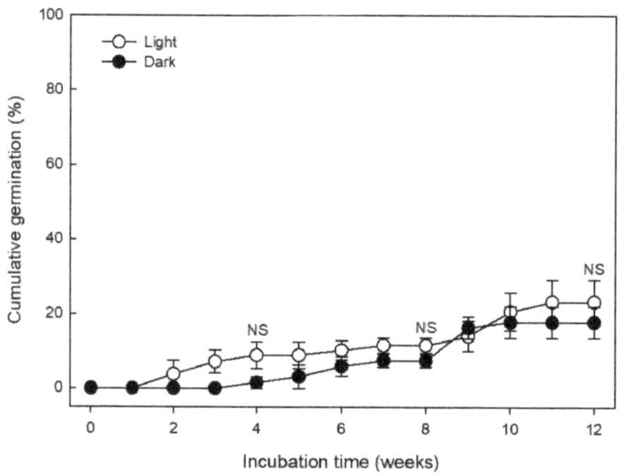

Figure 4. Germination of *Scutellaria indica* L. var. *coccinea* S.T.Kim & S.T.Lee seeds as affected by the performed light regimes. The seeds were incubated in light (12 h photoperiod) and darkness (24 h) regimes at 25/15 °C for 12 weeks. Error bars indicate mean ± standard error of four replicates of 20 seeds. NS means non-significant at $p < 0.05$ (Duncan's multiple range tests).

3.4.2. Effect of Temperature on Germination

The results showed that the seeds did not germinate under any of the temperature conditions except for 25/15 °C, after 4 weeks of incubation (Figure 5). At 25/15 °C, the seeds germinated after 3 weeks of incubation, with a rate of 1.7%, but there was no statistically significant difference when compared to the other temperature treatments. The final germination rates at 4, 15/6, 20/10, and 25/15 °C were 0.0, 0.0, 10.3, and 13.6%, respectively, but there were no statistically significant differences between the temperature treatments.

3.4.3. Effect of Cold Stratification on Germination

The cold stratification (CS) experiments were conducted twice. In both Experiments 1 and 2, the final germination increased as the CS period increased (Figure 6). In Experiment 1, the mean germination time (MGT) decreased from 61.5 d (CS for 0 weeks) to 12.8 d (CS for 8 weeks). For the control (CS for 0 weeks), the germination percentage did not reach 50%, and there was no significant difference in the time to reach 50% germination (T_{50}) values between the CS for 4 weeks and 8 weeks (Table 2). Similarly, in Experiment 2, the MGT decreased from 45.7 (control) d to 5.8 d. Additionally, T_{50} decreased from 10.5 (CS for 4 weeks) d to 2.6 (CS for 12 weeks), indicating that the extended the CS period, the shorter the time required for germination (Table 2). In Experiment 1, the germination index (GI) increased from 0.06 (control) to 1.48 (12 weeks of CS), and in Experiment 2, it increased from 0.14 (control) to 7.11 (12 weeks). The germination percentage was significantly affected by the cold stratification treatment, as indicated by the one-way ANOVA (Table 3).

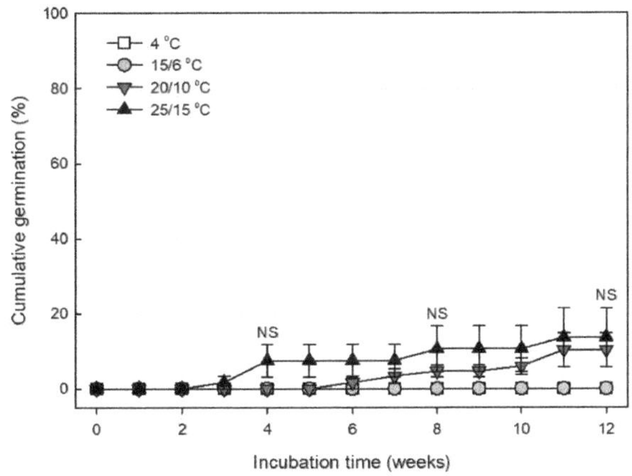

Figure 5. Germination of *Scutellaria indica* L. var. *coccinea* S.T.Kim & S.T.Lee seeds as affected by the performed temperature treatments. The seeds were incubated in light/dark (12 h photoperiod) at four temperature conditions (4, 15/6, 20/10, or 25/15 °C) for 12 weeks. Error bars indicate mean ± standard error of four replicates of 20 seeds. NS means non-significant at $p < 0.05$ (Duncan's multiple range tests).

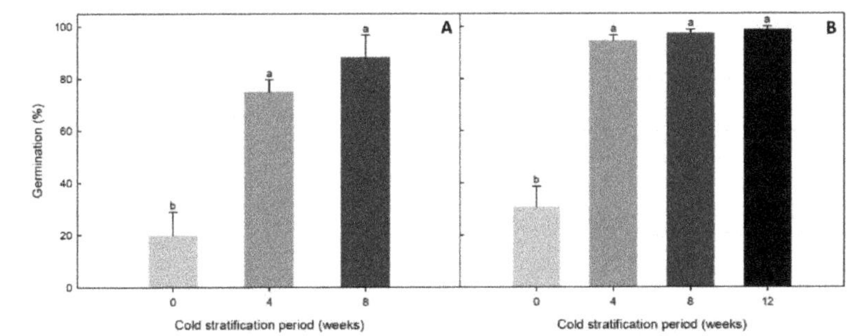

Figure 6. Germination of *Scutellaria indica* L. var. *coccinea* S.T.Kim & S.T.Lee seeds as affected by the cold stratification period (0, 4, 8, or 12 weeks) at 4 °C. After the treatment, the seeds were incubated at 25/15 °C and the final germination was investigated when there was no change in germination over 3 weeks; seeds were incubated for 12, 8, or 4 weeks (**A**) and 12, 8, 6, or 6 weeks (**B**), respectively. Error bars indicate mean ± standard error of four replicates of 20 seeds. The different lowercase letters represent significant difference at $p < 0.05$ (Duncan's multiple range tests).

Table 2. Mean germination time (MGT), time to reach 50% germination (T_{50}), and germination index (GI) of *Scutellaria indica* L. var. *coccinea* S.T.Kim & S.T.Lee seeds as affected by cold stratification at 4 °C for various periods. This experiment was conducted twice. In Experiment 1, the data were collected at 12, 8, or 4 weeks after incubation at 25/15 °C following cold stratification treatment for 0, 4, or 8 weeks, respectively. In Experiment 2, the data were collected at 12, 8, 6, or 6 weeks after incubation at 25/15 °C following cold stratification treatment for 0, 4, 8, or 12 weeks, respectively. Both experiments were ended when the germination rate showed no further changes for more than 3 weeks.

Experiment	Cold Stratification Period (Weeks)	MGT (Days)		T_{50} (Days)		GI	
	0	61.46 ± 6.33 [1]	a [2]	NG [3]		0.06 ± 0.01	b
1	4	11.80 ± 0.82	b	7.25 ± 1.88	a	1.38 ± 0.05	a
	8	12.79 ± 0.58	b	10.50 ± 0.00	a	1.48 ± 0.19	a

Table 2. *Cont.*

Experiment	Cold Stratification Period (Weeks)	MGT (Days)		T_{50} (Days)		GI	
2	0	45.68 ± 4.82	a	NG		0.14 ± 0.02	b
	4	15.32 ± 0.27	b	10.50 ± 0.00	a	1.34 ± 0.03	b
	8	11.88 ± 1.42	c	6.13 ± 1.68	b	2.08 ± 0.07	b
	12	5.79 ± 1.14	c	2.63 ± 0.88	c	7.11 ± 1.12	a

[1] Values indicate mean ± standard error (n = 4). [2] The different lowercase letters indicate significant differences ($p < 0.05$), Duncan's multiple-range test. [3] NG: Not germinated to 50%.

Table 3. One-way ANOVA was conducted to examine the effects of cold stratification (Experiment 1 and Experiment 2) on the final germination, mean germination time (MGT), T_{50} (time to reach 50% germination), and GI (germination index) and the effects of GA_3 on the final germination of the *Scutellaria indica* L. var. *coccinea* S.T.Kim & S.T.Lee seeds.

		DF	SS	MS	F-Value	*p*-Value
Cold stratification (Experiment 1)	Germination	2	10,556.198	5278.099	21.38	0.0004
	MGT	2	6438.010	3219.005	58.79	<0.0001
	T_{50}	1	21.125	21.125	3.00	0.1340
	GI	2	5.078	2.539	36.48	<0.0001
Cold stratification (Experiment 2)	Germination	3	13,181.046	4393.682	62.37	<0.0001
	MGT	3	3793.997	1264.666	47.56	<0.0001
	T_{50}	2	124.542	62.271	13.07	0.0022
	GI	3	112.995	37.665	9.16	0.0020
GA_3	Germination	3	33,398.152	11,132.717	358.33	<0.0001

3.4.4. Effect of GA_3 Treatment on Germination

Germination was promoted as the GA_3 concentration increased (Figure 7). In the second week of incubation, germination began to increase in the 100 and 1000 mg·L^{-1} GA_3 treatments. The final germination rates were the highest at 97.4 and 100.0%, respectively, and there was no significant difference between the two treatments. In the 10 mg·L^{-1} GA_3 treatment, similarly, germination started in the second week of incubation, but the germination was lower than the 100 and 1000 mg·L^{-1} GA_3, with a final germination of 17.8%, which was not different from the 0 mg·L^{-1} GA_3.

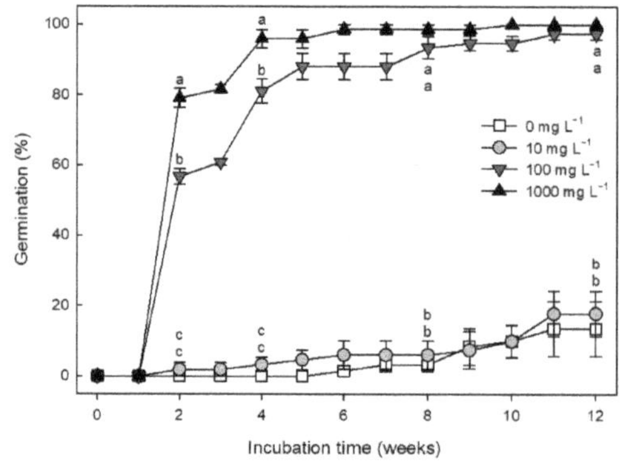

Figure 7. Germination of *Scutellaria indica* L. var. *coccinea* S.T.Kim & S.T.Lee seeds as affected by GA_3 (0, 10, 100, or 1000 mg·L^{-1}). The seeds treated with GA_3 for 24 h were incubated at 25/15 °C for 12 weeks. Error bars indicate mean ± standard error of four replicates of 20 seeds. The different letters indicate significant differences at $p < 0.05$ (Duncan's multiple range tests).

4. Discussion

Information on germination, bud formation, flowering, and seed development, obtained through phenological studies, is essential for maintaining a proper phenological cycle and ensuring the reproductive success of the species [54]. In our phenological experiments, seeds of *S. indica* var. *coccinea* were buried in the field in mid-July and remained ungerminated throughout the winter but began to germinate the following spring when daily soil temperatures reached approximately 2.9 °C (Figure 2A,B), indicating that germination can occur under relatively low temperatures. By contrast, under controlled laboratory conditions, the final germination under light and temperature treatments was only about 20% (Figures 4 and 5), whereas cold stratification (CS) treatment significantly increased germination to more than 90% (Figure 6). Collectively, these findings indicate that freshly matured seeds of *S. indica* var. *coccinea* are unable to germinate immediately after dispersal and instead require a period of dormancy release prior to successful germination. The consistent outcomes obtained from both field and laboratory experiments provide strong evidence that this species exhibits physiological dormancy. Therefore, the ecological significance of our results lies in demonstrating that exposure to a natural cold period is an important factor in breaking dormancy, which is supported by laboratory studies showing that CS promotes germination.

Baskin and Baskin classified seed dormancy into five classes: physiological dormancy (PD), morphological dormancy (MD), morphophysiological dormancy (MPD), physical dormancy (PY), and combined dormancy [54]. PD is the most common type of dormancy on Earth. Seeds with PD have a fully developed embryo at the time of dispersal from the mother plant; however, their germination is delayed by more than 30 d, even under favorable environmental conditions for germination. Several past studies have reported that many species belonging to the Lamiaceae family exhibit PD or nondormancy (ND). Especially in temperate regions, seeds of many species exhibit PD of varying depths, which gradually break during winter and germinate the following spring [40,55,56]. Seeds with MD had underdeveloped embryos at the time of dispersal from the mother plant. Once the embryo has grown sufficiently, the seeds can germinate within 30 days under favorable environmental conditions, without the need for any special treatments. MPD is a combination of the characteristics of PD and MD, and seeds with MPD require a longer period of embryonic growth and germination than those with MD. Seeds exhibiting PY can delay germination due to the presence of water-impermeable cell layers in the seed coat or fruit coat. Seeds with combinational dormancy combine the characteristics of PY and PD and have both a water-impermeable seed coat and a physiologically dormant embryo [42,57,58]. For example, *Teucrium santae* of the Lamiaceae family is known to exhibit this type of dormancy. Physical dormancy could be broken through chemical scarification with sulfuric acid, and physiological dormancy could be overcome with 1500 mg·L^{-1} GA for 72 h [47].

Mature seeds of *S. indica* var. *coccinea* embryos are fully developed at the time of dispersal from the mother plant (Figure 1D). This indicates that the seeds do not require additional time for embryo growth and therefore do not exhibit MD or MPD. In the water absorption experiment conducted to determine the permeability of the seed coat, seed mass increased by more than 20% within 6 h (Figure 3). This indicates that the seed coat of *S. indica* var. *coccinea* is permeable, and therefore, the seeds do not exhibit PY or combinational dormancy [59]. Based on our light and temperature treatment experiments, the majority of seeds showed a delay in germination of more than 30 d, even under favorable environmental conditions (Figures 4 and 5). Therefore, mature seeds of *S. indica* var. *coccinea* can be considered to exhibit PD, as they do not germinate due to the low growth potential of the embryo [42].

PD is classified into three levels (non-deep, intermediate, and deep) based on dormancy depth [60,61]. This classification is determined by the seed response to treatments, such as cold or warm stratification, gibberellic acid (GA_3), and dry after-ripening [60,62]. Non-deep PD is broken down by cold or warm stratification and gibberellic acid (GA_3) treatment. Intermediate PD showed varying responses to GA treatment depending on the species, and dormancy was disrupted by CS for 2–3 months. Deep PD does not respond to GA_3 treatment and requires CS for more than three months to break dormancy [42]. The performed CS experiments were conducted twice (Experiments 1 and 2). In both Experiments 1 and 2, CS for more than four weeks was effective in reducing the dormancy depth of *S. indica* var. *coccinea* seeds and promoted germination (Tables 2 and 3 and Figure 6). As a result of cold stratification at 4 °C, in Experiment 1, as the CS period expanded, mean germination time (MGT) shortened from an average of 61.5 d to 12.8 d, and time to reach 50% germination (T_{50}) showed no significant difference. In Experiment 2, as the CS period increased, MGT decreased from an average of 45.7 d to 5.8 d, and T_{50} shortened from 10.5 d to 2.6 d. As a result of the GA_3 treatment experiment, germination was promoted with increasing concentrations of GA_3. Similar results were observed for the germination characteristics of two *Polygonella* species, which have non-deep PD of seeds by Heather et al. [63]. In that study, *P. polygama* seeds treated with 1000 $mg \cdot L^{-1}$ GA_3 reached a final germination of 32.5%, which was about double that of the control. Many previous studies have demonstrated the positive effects of gibberellic acid on seed germination in species belonging to the Lamiaceae family. For example, Szekely-Varga et al. [64] reported that two *Lavandula angustifolia* cultivars (Codreanca and Sevtopolis) recorded germination of over 90% with 300 $mg \cdot L^{-1}$ GA_3, approximately 2.5 times higher than the control. In addition, the seeds of photoblastic species *Nepeta rtanjensis* showed enhanced germination under dark conditions when treated with approximately 350 $mg \cdot L^{-1}$ GA_3, indicating that GA_3 was able to fully substitute the light requirement for seed germination in *N. rtanjensis* [65]. Based on our results, we classified the seeds of *S. indica* var. *coccinea* as having a non-deep form of PD since germination was promoted by a relatively short period of CS and GA_3 treatment. Similar effects have been observed in temperate montane wetland plants in China, including *Betula ovalifolia*, *Carex limosa*, *Hypericum longisylum*, and *Lobelia sessilifolia*. When the seed germination characteristics of the four plant species and their dormancy types were investigated, it was found that germination was promoted through CS and that germination was effectively promoted by inducing mechanical damage to the seed coat through sulfuric acid pre-treatment and removing potential PD through GA_3 treatment [66]. In conclusion, all four species exhibited non-deep PD, supporting the claim that PD is a common adaptive strategy in plants native to temperate regions.

Despite the wide geographical distribution of plants in the genus *Scutellaria*, many species in this genus are known to exhibit similar classes of seed dormancy (PD or ND) [42,55]. This suggests that seed dormancy traits inherited from their common ancestor have been evolutionarily conserved over an extended period. This phenomenon can be interpreted as a representative example of trait stasis in which certain physiological and ecological traits remain almost unchanged over long evolutionary periods [43]. Physiological traits are regulated by genes; therefore, species belonging to the same family or genus are likely to exhibit similar characteristics, such as seed germination [67]. Although the level of PD has not been classified, previous studies on closely related species have shown similar effects, where dormancy was broken and germination was promoted by CS. According to Liu et al. [67] and Baskin and Baskin [40], seeds of *Ajuga lupulina* and *Scutellaria baicalensis*, which belong to the Lamiaceae family, exhibit PD, and CS is effective in promoting germination under dry conditions [40,67]. In addition, the PD of *Agstache foeniculum* seeds is broken through CS, and in the case of *Amthystea caerulea* seeds, the PD can be broken by 40 d of CS. In contrast, in some species such as *Scutellaria parvula*, which belongs to the same genus, dormancy is broken by high summer temperatures [44,68], thus indicating that germination can vary slightly even among species within the same genus [40,62]. These differences in environmental responses to dormancy breaking may be the result of trait divergence, in which the basic physiological traits are preserved, but responsiveness varies according to the climatic conditions [43,69]. Therefore, the genus *Scutellaria* can be regarded as an example in which traits of stasis and divergence coexist. Therefore, understanding the germination requirements of seeds during dormancy breaking and induction is essential, as they play a crucial role in the control of germination timing, seedling establishment, and survival [70].

Based on our experiment results, we classified the seeds of *S. indica* var. *coccinea* as exhibiting non-deep PD. It is a common type found in many herbaceous plants and some woody species, and it has been estimated that more than 90% of seeds exhibiting PD have this level of PD [71]. There are six types of non-deep PD that are further classified based on the dormancy-breaking process and the temperature requirements for germination after dormancy have been broken [40,42,72]. When seed germination is restricted by specific environmental factors, resulting in conditional dormancy (CD) within a narrow range of conditions, it is classified as type 1, 2, or 3. If the seeds do not exhibit CD, they are classified as types 4, 5, or 6 [40,42,70]. Under field conditions, the *S. indica* var. *coccinea* germinated at a low rate of 8.1%, whereas most of the seeds remained in a dormant state. By mid-July of the following year, all seeds had germinated, suggesting that *S. indica* var. *coccinea* remained in a dormant (D) state and gradually transitioned from a dormant through a CD state to a non-dormant (ND) state as dormancy was broken by exposure to low temperatures. This pattern (D $\leftrightarrow$ ND) reflects the typical dormancy cycle observed in non-deep PD species native to temperate regions [73]. Based on these aforementioned observations, we demonstrate that the seeds of *S. indica* var. *coccinea* exhibit CD; therefore, we can narrow the non-deep PD types of the target species to types 1, 2, and 3. According to Baskin and Baskin (2014) and Soltani et al. (2017), species with type 1 non-deep PD initially germinate only at low temperatures, and as dormancy breaks down, the maximum germination temperature increases [40,70]. Seeds with type 2 non-deep PD initially germinate only at high temperatures, and as dormancy is broken, the germinable minimum temperature decreases, whereas seeds with type 3 non-deep PD initially germinate only at intermediate temperatures, and as dormancy is broken, both the lower and upper temperature limits for germination expand. Mature seeds of *S. indica* var. *coccinea* can germinate initially at a low rate only under the high temperatures ($\geq$20 °C) of summer (Figures 2 and 5). However, increased germination with the CS treatment indicated that the temperature range for germination expanded to lower temperatures as the seeds were exposed to low temperatures (Figure 6). If the temperature range for seed germination is broadened, seeds may become less dependent on specific climatic conditions and thus be able to germinate in a wider range of environments. Therefore, the range of favorable conditions for both survival and reproduction is expanded, thus allowing for greater reproductive success. This may be a common characteristic in regions with large temperature fluctuations and may play an important role in maintaining species reproduction and diversity in response to climate change [74,75]. In fact, under field conditions, the first day on which the germination of *S. indica* var. *coccinea* seeds exceeded 50% was in early March, 35 weeks after being buried, when the daily average soil temperature was 9.6 °C (Figure 2). This indicates that low temperatures are the primary environmental signals for breaking dormancy in *S. indica* var. *coccinea*, and the minimum temperature required for germination decreased. Therefore, it can be concluded that the seeds of *S. indica* var. *coccinea* exhibited type 2 non-deep PD (Table 4). This type is commonly found in the seeds of summer annual plants and many perennial plants in temperate regions [70,73,76]. In temperate regions with distinct seasonal changes, such as Korea, plant survival and reproduction depend on germination strategies that are well-adapted to the environment [77–79]. By maintaining dormancy from dispersal to winter and by germinating in early spring, *S. indica* var. *coccinea* uses a strategy that selects a favorable period for survival. This characteristic may not merely be an individual species response but rather a common evolutionary trait shared among the genus *Scutellaria* in temperate climates. The ecological traits of perennial herbaceous plants native to temperate forests in eastern Asia and eastern North America remain in a state of evolutionary stasis and maintain stable ecological requirements [80,81]. This supports the results of the present study, thus indicating that the genus *Scutellaria* has maintained germination traits adapted to temperate climates over long evolutionary periods. In this way, the unique germination traits of each species not only help maintain genetic diversity but also serve as a strategy for adapting more effectively to changing environmental conditions [82,83].

Table 4. Evaluation and summary of seed dormancy and germination traits in *Scutellaria indica* var. *coccinea* based on Baskin & Baskin's criteria for non-deep physiological dormancy.

Seed Traits	Observation in *S. indica* var. *coccinea*
Embryo development	Embryo was foliate, bent, and fully developed (no morphological dormancy)
Water absorption	Permeable seed coat (no physical and combinational dormancy)
Effect of temperature regimes on germination	≤20% germination (seeds were dormant)
Effect of cold stratification on germination	4 °C, ≥8weeks → ≥80% germination at 25/15 °C
Effect of GA_3 treatment on germination	GA_3 100, 1000 mg·L^{-1} → ≥80% germination at 25/15 °C
Germination phenology	After exposure to cold winter temperatures, spring germination at low soil temperatures (≤10 °C) (temperature range at which seeds can germinate gradually increases from high to low)
Dormancy type	Type 2 non-deep physiological dormancy [1]

[1] Type 2 non-deep PD was classified according to the references [40,42].

Such survival strategies have also been observed in the life cycles of plants transplanted into the field. After the plants matured, they flowered in spring and produced seeds (Figure 1H). The fresh seeds of *S. indica* var. *coccinea* dispersed in summer exhibited dormancy (Figure 1I), with most remaining dormant in the soil, thereby forming a soil seed bank. Soil seed banks are broadly categorized into transient and persistent types [84]. A transient seed bank consists of seeds that do not remain viable in the soil for more than one year before germination, whereas a persistent seed bank comprises seeds that remain dormant in the soil for at least one year before germination [85,86]. Transient seed banks can be classified into two types. Type 1 transient seed banks involve seeds that undergo dormancy release in late spring and summer and typically germinate in autumn. In contrast, type 2 transient seed banks consist of seeds that experience dormancy release during winter and germinate in the early spring of the following year [84,87]. This type of seed bank is commonly observed in species inhabiting northern and continental temperate regions [84,88]. Our observations indicate that the seeds of *S. indica* var. *coccinea* exhibit the characteristics of a type 2 transient seed bank, undergoing dormancy release during winter and germinating the following spring. This suggests that *S. indica* var. *coccinea* has developed a type 2 transient seed bank as an adaptation to the geographical features of temperate regions. Through the repetition of their life cycle, the plants can adapt to seasonal changes and environmental factors, thereby maintaining the diversity of their clusters [70,89]. The results of the present study provide important baseline data for the mass propagation of *S. indica* var. *coccinea*.

In our phenology experiment, the seedling emergence percentage (73.7%) was lower than the germination percentage (100.0%), indicating that not all germinated seeds successfully developed into seedlings. Similar discrepancies have been reported in other species, as seed germination does not necessarily ensure successful recruitment due to intrinsic and extrinsic factors [75,90]. Post-germination mortality can be caused by fungal pathogens such as Pythium and Rhizoctonia, abiotic stressors like insufficient precipitation and soil desiccation, or other environmental constraints [91,92]. Most of the deceased individuals were either dry or had rotted with a dark black seed coat. Since our observations were made under natural field conditions, the precise causes of mortality between germination and emergence remain unclear. Nevertheless, this discrepancy highlights the importance of monitoring both germination and seedling establishment when evaluating recruitment dynamics. This study confirmed that strategies such as prescreening seed quality, maintaining adequate soil moisture, applying pathogen control, and minimizing predation risk are essential to reduce post-germination mortality and enhance propagation efficiency.

Practical Conservation Applications

For *S. indica* var. *coccinea*, the results of this study provide a practical propagation protocol that can be directly applied to conservation programs. Dormancy can be effectively relieved by cold stratification of seeds at 4 °C for 4 weeks or by treating them with 100 $mg{\cdot}L^{-1}$ GA_3, a cost-effective alternative to the commonly used 1000 $mg{\cdot}L^{-1}$ concentration. Once dormancy is relieved, optimal germination occurs under light conditions at temperatures ranging from 20/10 °C to 25/15 °C. For successful field establishment, transplanting seedlings is effective when soil temperatures rise above approximately 10 °C. Careful monitoring of post-germination mortality during the early stages of seedling development is also important, as this can significantly impact establishment success. These recommendations provide a concise and practical framework that enhances the applicability of this study to conservation professionals.

5. Conclusions

Based on the experimental results of the various treatments on the germination characteristics of *S. indica* var. *coccinea* seeds were classified as exhibiting type 2 non-deep PD. Freshly mature seeds contain fully developed embryos that remain dormant. In the field, dormancy was released during winter and germinated the following spring. In laboratory experiments, cold stratification (CS) for more than 4 weeks effectively broke dormancy, thus resulting in a germination rate of nearly 100.0%. These results confirmed that low temperature is a crucial environmental factor in breaking the PD of *S. indica* var. *coccinea* seeds, and the minimum temperature required for germination increased as dormancy was released. GA_3 was able to substitute for the requirements of low temperatures, and germination was promoted at concentrations of 100 and 1000 $mg{\cdot}L^{-1}$, allowing all of the seeds to germinate. Considering the efficiency, time, and cost, we suggest breaking seed dormancy and promoting propagation through either CS for 4 weeks or 100 $mg{\cdot}L^{-1}$ of GA_3 treatment. The results of this study provide fundamental information for developing mass propagation techniques for *S. indica* var. *coccinea* and can serve as a valuable resource for ex situ conservation efforts in botanical gardens.

Author Contributions: Conceptualization, H.C.K. and S.Y.L.; Investigation and Formal analysis, J.H.K. and S.Y.L.; Data curation, J.H.K.; Funding acquisition, H.C.K. and S.Y.L.; Methodology, Supervision, an Project administration, S.Y.L.; Writing—original draft preparation, J.H.K. and S.Y.L.; Writing—review and editing, H.C.K. and S.Y.L. All authors have read and agreed to the published version of the manuscript.

Funding: This work was supported by the Korea Institute of Science and Technology (KIST) ORP program [BlueBell Research Project (grant number 2E33511-25-022)] and the Basic Science Research Program through the National Research Foundation of Korea (NRF) funded by the Ministry of Education (grant number NRF-2018R1A6A1A0302482).

Data Availability Statement: The original contributions presented in this study are included in the article. Further inquiries can be directed to the corresponding author.

Acknowledgments: Thanks to every member of the Floriculture and Landscape Plants Lab (FLPL, Gyeongkuk National University).

Conflicts of Interest: The authors declare no conflicts of interest.

References

1. Noss, R.F. Indicators for monitoring biodiversity: A hierarchical approach. *Conserv. Biol.* **1990**, *4*, 355–364. [CrossRef]
2. Redford, K.H.; Richter, B.D. Conservation of biodiversity in a world of use. *Conserv. Biol.* **1999**, *13*, 1246–1256. [CrossRef]
3. Ramanatha Rao, V.; Hodgkin, T. Genetic diversity and conservation and utilization of plant genetic resources. *Plant Cell Tissue Organ Cult.* **2002**, *68*, 1–19. [CrossRef]
4. Gitay, H.; Suárez, A.; Watson, R.T.; Dokken, D.J. *Climate Change and Biodiversity (IPCC Technical Paper V)*; Intergovernmental Panel on Climate Change: Geneva, Switzerland, 2002.
5. Bélanger, J.; Pilling, D. *The State of the World's Biodiversity for Food and Agriculture*; Food and Agriculture Organization of the United Nations (FAO): Rome, Italy, 2019.
6. McNeely, J.A. Lessons from the past: Forests and biodiversity. *Biodivers. Conserv.* **1994**, *3*, 3–20. [CrossRef]
7. McNeely, J.A.; Gadgil, M.; Leveque, C.; Padoch, C.; Redford, K. Human influence in biodiversity. In *Global Biodiversity Assessment*; Heywood, V.H., Ed.; Cambridge University Press: Cambridge, UK, 1995.
8. Sodhi, N.S.; Ehrlich, P.R. (Eds.) *Conservation Biology for All*; Oxford University Press: Oxford, UK, 2010.
9. Pyšek, P.; Jarošík, V.; Hulme, P.E.; Pergl, J.; Hejda, M.; Schaffner, U.; Vilà, M. A global assessment of invasive plant impacts on resident species, communities and ecosystems. *Glob. Change Biol.* **2012**, *18*, 1725–1737. [CrossRef]
10. Corlett, R.T. Plant diversity in a changing world: Status, trends, and conservation needs. *Plant Divers.* **2016**, *38*, 10–16. [CrossRef]
11. Kim, Y.S. Current Status and Development Strategies of Korean Botanical Gardens. *Proc. Korean Soc. Environ. Ecol. Conf.* **2005**, *15*, 6–11.
12. Kim, Y.S. Conservation of plant diversity in Korea. *Landsc. Ecol. Eng.* **2006**, *2*, 163–170. [CrossRef]
13. Lee, B.H. Destruction of nature and the crisis of biodiversity (Jayeon pahoe wa saengmul dayangseong-ui wigi). *Sci. Technol.* **1991**, *24*, 42–47.
14. Hong, H.J.; Kim, C.K.; Lee, H.W.; Lee, W.K. Conservation, restoration, and sustainable use of biodiversity based on habitat quality monitoring: A case study on Jeju Island, South Korea (1989–2019). *Land* **2021**, *10*, 774. [CrossRef]
15. Anderson, S. Area and endemism. *Q. Rev. Biol.* **1994**, *69*, 451–471. [CrossRef]
16. Primack, R.B. *A Primer of Conservation Biology*, 5th ed.; Sinauer Associates: Sunderland, MA, USA, 2008.
17. Burlakova, L.E.; Karatayev, A.Y.; Karatayev, V.A.; May, M.E.; Bennett, D.L.; Cook, M.J. Endemic species: Contribution to community uniqueness, effect of habitat alteration, and conservation priorities. *Biol. Conserv.* **2011**, *144*, 155–165. [CrossRef]
18. Chung, G.Y.; Chang, K.S.; Chung, J.M.; Choi, H.J.; Paik, W.K.; Hyun, J.O. A checklist of endemic plants on the Korean Peninsula. *Korean J. Plant Taxon.* **2017**, *47*, 264–288. [CrossRef]
19. Cho, Y.C.; Seol, J.; Lim, C.H. Climate-induced distribution dynamics and niche adaptation of South Korean endemic plants across the Korean Peninsula. *Sci. Rep.* **2024**, *14*, 22253. [CrossRef]
20. Heywood, V.H. The future of plant conservation and the role of botanic gardens. *Plant Divers.* **2017**, *39*, 309–313. [CrossRef]
21. Shin, H.H.; Kang, S.K. The trends in the application of the Global Strategy for Plant Conservation to arboretums and countermeasures. *J. Korean Inst. Forest Recreat.* **2011**, *15*, 91–102.
22. Choi, S.R. Content Analysis and Recognition Survey on Biodiversity Education Programs in National Institute of Biological Resources. Master's Thesis, Seoul National University, Seoul, Republic of Korea, 2018.

23. Havens, K.; Vitt, P.; Maunder, M.; Guerrant, E.O.; Dixon, K. Ex situ plant conservation and beyond. *BioScience* **2006**, *56*, 525–531. [CrossRef]
24. Paton, A. A global taxonomic investigation of *Scutellaria* (Labiatae). *Kew Bull.* **1990**, *45*, 399–450. [CrossRef]
25. Lee, Y.; Kim, S. The complete chloroplast genome of *Scutellaria indica* var. *coccinea* (Lamiaceae), an endemic taxon in Korea. *Mitochondrial DNA B Resour.* **2019**, *4*, 2539–2540. [CrossRef]
26. Kim, S.T.; Lee, S.T. Two new varieties of *Scutellaria indica* L. from Korea. *Korean J. Plant Taxon.* **1994**, *24*, 73–78. [CrossRef]
27. Nishikawa, K.; Furukawa, H.; Fujioka, T.; Fujii, H.; Mihashi, K.; Shimomura, K.; Ishimaru, K. Phenolics in tissue cultures of *Scutellaria*. *Recent Res. Dev. Phytochem.* **2000**, *4*, 55–60.
28. Guo, X.R.; Wang, X.G.; Su, W.H.; Zhang, G.F.; Zhou, R. DNA barcodes for discriminating the medicinal plant *Scutellaria baicalensis* (Lamiaceae) and its adulterants. *Biol. Pharm. Bull.* **2011**, *34*, 1198–1203. [CrossRef]
29. Joshee, N.; Tascan, A.; Medina-Bolivar, F.; Parajuli, P.; Rimando, A.M.; Shannon, D.A.; Adelberg, J.W. *Scutellaria*: Biotechnology, phytochemistry and its potential as a commercial medicinal crop. In *Biotechnology for Medicinal Plants: Micropropagation and Improvement*; Chandra, S., Lata, H., Varma, A., Eds.; Springer: Berlin/Heidelberg, Germany, 2013; pp. 69–99.
30. Cole, I.B.; Saxena, P.K.; Murch, S.J. Medicinal biotechnology in the genus *Scutellaria*. *Vitro Cell. Dev. Biol. Plant* **2007**, *43*, 318–327. [CrossRef]
31. Shang, X.; He, X.; He, X.; Li, M.; Zhang, R.; Fan, P.; Zhang, Q.; Jia, Z. The genus *Scutellaria*: An ethnopharmacological and phytochemical review. *J. Ethnopharmacol.* **2010**, *128*, 279–313. [CrossRef]
32. Wang, Z.L.; Wang, S.; Kuang, Y.; Hu, Z.M.; Qiao, X.; Ye, M. A comprehensive review on phytochemistry, pharmacology, and flavonoid biosynthesis of *Scutellaria baicalensis*. *Pharm. Biol.* **2018**, *56*, 465–484. [CrossRef]
33. Korea Heritage Service. Sangumburi Crater. Jeju. Available online: https://www.heritage.go.kr/main/?v=1756202055699 (accessed on 6 April 2025).
34. Korea Institute of Oriental Medicine (KIOM). *Scutellaria indica* var. *coccinea*. Available online: https://www.kmif.org/board/board.php?b_id=pds (accessed on 6 April 2025).
35. Korea National Arboretum (KNA). *Scutellaria indica* var. *coccinea*. Available online: https://terms.naver.com/entry.naver?docId=3541232&cid=46694&categoryId=46694 (accessed on 6 April 2025).
36. National Institute of Biological Resources (NIBR). Available online: https://species.nibr.go.kr/home/mainHome.do (accessed on 5 June 2025).
37. Korea National Arboretum (KNA). *Establishment of Conservation and Restoration Infrastructure for Rare Endemic Plants*; Reported No. TRKO201500014027; Korea Forest Service: Daejeon, Republic of Korea, 2014.
38. Abeli, T.; Dalrymple, S.; Godefroid, S.; Mondoni, A.; Müller, J.V.; Rossi, G.; Orsenigo, S. Ex situ collections and their potential for the restoration of extinct plants. *Conserv. Biol.* **2020**, *34*, 303–313. [CrossRef]
39. Doijode, S.D. Seed conservation-An effective method for preservation of genetic variability in French bean (*Phaseolus vulgaris* L.). *Indian J. Plant Genet. Resour.* **2001**, *14*, 290–292.
40. Baskin, C.C.; Baskin, J.M. *Seeds: Ecology, Biogeography and Evolution of Dormancy and Germination*, 2nd ed.; Academic Press: Cambridge, MA, USA; Elsevier: Amsterdam, The Netherlands, 2014.
41. Bewley, J.D. Seed germination and dormancy. *Plant Cell* **1997**, *9*, 1055–1066. [CrossRef]
42. Baskin, J.M.; Baskin, C.C. A classification system for seed dormancy. *Seed Sci. Res.* **2004**, *14*, 1–16. [CrossRef]
43. Adams, C.A.; Baskin, J.M.; Baskin, C.C. Trait stasis versus adaptation in disjunct relict species: Evolutionary changes in seed dormancy-breaking and germination requirements in a subclade of *Aristolochia subgenus* Siphisia (Piperales). *Seed Sci. Res.* **2005**, *15*, 161–173. [CrossRef]
44. Baskin, J.M.; Baskin, C.C. Ecological life cycle and temperature relations of seed germination and bud growth of *Scutellaria* parvula. *Bull. Torrey Bot. Club.* **1982**, *109*, 1–6. [CrossRef]
45. Yoshimura, H.; Arakaki, S.; Hamagawa, M.; Kitamura, Y.; Yokota, M.; Denda, T. Differentiation of germination characteristics in *Scutellaria* rubropunctata (Lamiaceae) associated with adaptation to rheophytic habitats in the subtropical Ryukyu Islands of Japan. *J. Plant Res.* **2019**, *132*, 359–368. [CrossRef]
46. Gorai, M.; Gasmi, H.; Neffati, M. Factors influencing seed germination of medicinal plant *Salvia aegyptiaca* L. (Lamiaceae). *Saudi J. Biol. Sci.* **2011**, *18*, 255–260. [CrossRef]
47. Bentekhici, M.; Mehdadi, Z.; Latreche, A. Seed germination behavior of *Teucrium santae* Quézel & Simonneau: A vulnerable and endemic Lamiaceae (Northwest Algeria). *Folia Oecol.* **2023**, *50*, 174–184.
48. Oliveira, M.F.; Cruz, C.R.P.; de Oliveira, L.M. Morphometric characterization of the seeds and germination of six endemic *Lamiaceae* species in northeastern Brazil. *Braz. J. Agric.* **2018**, *93*, 69–79.
49. Matthews, S.; Khajeh-Hosseini, M. Mean germination time as an indicator of emergence performance in soil of seed lots of maize (*Zea mays*). *Seed Sci. Technol.* **2006**, *34*, 339–347. [CrossRef]
50. Matthews, S.; Khajeh-Hosseini, M. Length of the lag period of germination and metabolic repair explain vigour differences in seed lots of maize (*Zea mays*). *Seed Sci. Technol.* **2007**, *35*, 200–212. [CrossRef]

51. Mavi, K.; Demir, I.; Matthews, S. Mean germination time estimates the relative emergence of seed lots of three cucurbit crop under stress conditions. *Seed Sci. Technol.* **2010**, *38*, 14–25. [CrossRef]
52. Scott, S.J.; Jones, R.A.; Williams, W.A. Review of data analysis methods for seed germination. *Crop Sci.* **1984**, *24*, 1192–119 [CrossRef]
53. Salehzade, H.; Shishvan, M.I.; Ghiyasi, M.; Forouzin, F.; Siyahjani, A.A. Effect of seed priming on germination and seedlin growth of wheat (*Triticum aestivum* L.). *Res. J. Biol. Sci.* **2009**, *4*, 629–631.
54. Miller, S.A.; Bartow, A.; Gisler, M.; Ward, K.; Young, A.S.; Kaye, T.N. Can an ecoregion serve as a seed transfer zone? Evidenc from a common garden study with five native species. *Restor. Ecol.* **2011**, *19*, 268–276. [CrossRef]
55. Martin, A.C. The comparative internal morphology of seeds. *Am. Midl. Nat.* **1946**, *36*, 513–660. [CrossRef]
56. An, K.; Yang, M.; Baskin, C.C.; Li, M.; Zhu, M.; Jiao, C.; Wu, H.; Zhang, P. Type 2 nondeep physiological dormancy in seeds o *Fraxinus chinensis* subsp. *rhynchophylla* (Hance) A.E. Murray. *Forests* **2022**, *13*, 1951. [CrossRef]
57. Baskin, C.C.; Baskin, J.M. When breaking seed dormancy is a problem: Try a move-along experiment. *Nat. Plants J.* **2003**, *4*, 17–2 [CrossRef]
58. Baskin, C.C.; Baskin, J.M. The natural history of soil seed banks of arable land. *Weed Sci.* **2006**, *54*, 549–557. [CrossRef]
59. Baskin, J.M.; Baskin, C.C. Classification, biogeography, and phylogenetic relationships of seed dormancy. In *Seed Conservation Turning Science into Practice*; Smith, R.D., Dickie, J.B., Linington, S.H., Pritchard, H.W., Probert, R.J., Eds.; Royal Botanic Gardens Kew, UK, 2003; pp. 518–544.
60. Nikolaeva, M.G. Factors controlling the seed dormancy pattern. In *The Physiology and Biochemistry of Seed Dormancy an Germination*; Khan, A.A., Ed.; North-Holland Publishing Company: Amsterdam, The Netherland, 1977; pp. 52–74.
61. Baskin, J.M.; Nan, X.; Baskin, C.C. A comparative study of seed dormancy and germination in an annual and a perennial specie of *Senna* (Fabaceae). *Seed Sci. Res.* **1998**, *8*, 501–512. [CrossRef]
62. Baskin, C.C.; Baskin, J.M. *Seeds: Ecology, Biogeography, and Evolution of Dormancy and Germination*; Academic Press: Cambridge MA, USA, 1998.
63. Heather, A.E.; Pérez, H.E.; Wilson, S.B. Non-deep physiological dormancy in seeds of two Polygonella species with horticultura potential. *HortScience* **2010**, *45*, 1854–1858. [CrossRef]
64. Szekely-Varga, Z.; Kentelky, E.; Cantor, M. Effect of gibberellic acid on the seed germination of *Lavandula angustifolia* Mill. *Rom. Hortic.* **2021**, *2*, 169–176. [CrossRef]
65. Todorović, S.; Živković, S.; Giba, Z.; Grubišić, D.; Mišić, D. Basic seed germination characteristics of the endemic species *Nepet rtanjensis* (Lamiaceae). *Plant Species Biol.* **2007**, *22*, 205–210. [CrossRef]
66. Wang, J.Y.; Bu, Z.J.; Poschlod, P.; Yusup, S.; Zhang, J.Q.; Zhang, Z.X. Seed dormancy types and germination response of 15 plan species in temperate montane peatlands. *Ecol. Evol.* **2024**, *14*, e11671. [CrossRef]
67. Liu, K.; Baskin, J.M.; Baskin, C.C.; Bu, H.; Liu, M.; Liu, W.; Du, G. Effect of storage conditions on germination of seeds of 48 species from high elevation grasslands of the eastern Tibet Plateau. *Am. J. Bot.* **2011**, *98*, 12–19. [CrossRef]
68. Baskin, C.C.; Baskin, J.M. Germination ecophysiology of herbaceous plant species in a temperate region. *Am. J. Bot.* **1988**, *75* 286–305. [CrossRef]
69. Grime, J.P. Trait convergence and trait divergence in herbaceous plant communities: Mechanisms and consequences. *J. Veg. Sc* **2006**, *17*, 255–260. [CrossRef]
70. Soltani, E.; Baskin, C.C.; Baskin, J.M. A graphical method for identifying the six types of non-deep physiological dormancy ir seeds. *Plant Biol.* **2017**, *19*, 673–682. [CrossRef]
71. Baskin, C.C.; Baskin, J.M. Seed dormancy in wild flowers. In *Flower Seeds: Biology and Technology*; McDonald, M.B., Kwong, F.Y. Eds.; CABI Publishing: Wallingford, UK, 2005; pp. 163–185.
72. Nur, M.; Baskin, C.C.; Lu, J.J.; Tan, D.Y.; Baskin, J.M. A new type of non-deep physiological dormancy: Evidence from thre annual Asteraceae species. *Seed Sci. Res.* **2014**, *24*, 301–314. [CrossRef]
73. Baskin, J.M.; Baskin, C.C. The annual dormancy cycle in buried weed seeds: A continuum. *BioScience* **1985**, *35*, 492–498. [CrossRef
74. Brändle, M.; Stadler, J.; Klotz, S.; Brandl, R. Distributional range size of weedy plant species is correlated to germination patterns *Ecology* **2003**, *84*, 136–144. [CrossRef]
75. Donohue, K.; Rubio de Casas, R.R.; Burghardt, L.; Kovach, K.; Willis, C.G. Germination, postgermination adaptation, and specie ecological ranges. *Annu. Rev. Ecol. Evol. Syst.* **2010**, *41*, 293–319. [CrossRef]
76. Baskin, C.C.; Baskin, J.M. Cold stratification in winter is more than enough for seed dormancy-break of summer annuals. *Seed Sc Res.* **2022**, *32*, 63–69. [CrossRef]
77. Archibold, O.W. *Ecology of World Vegetation*; Springer Science and Business Media: New York, NY, USA, 2012.
78. Gilliam, F.S. Forest ecosystems of temperate climatic regions: From ancient use to climate change. *New Phytol.* **2016**, *212*, 871–887 [CrossRef]
79. Kim, J.H. Analysis of Plant Phenological Changes in Temperate Forests of Korea. Master's Thesis, Seoul National University Seoul, Republic of Korea, 2019.

80. Ricklefs, R.E.; Latham, R.E. Intercontinental correlation of geographical ranges suggests stasis in ecological traits. *Am. Nat.* **1992**, *139*, 1305–1321. [CrossRef]
81. Qian, H.; Ricklefs, R.E. Geographical distribution and ecological conservatism of disjunct genera. *J. Ecol.* **2004**, *92*, 253–265. [CrossRef]
82. Angevine, M.W.; Chabot, B.F. Seed germination syndromes in higher plants. In *Topics in Plant Population Biology*; Solbrig, O.T., Jain, S., Johnson, G.B., Raven, P.H., Eds.; Columbia University Press: New York, NY, USA, 1979; pp. 188–206.
83. Bewley, J.D.; Black, M. 1. Development, Germination, and Growth. In *Physiology and Biochemistry of Seeds in Relation to Germination*; Springer: Berlin/Heidelberg, Germany, 2014.
84. Thompson, K.; Grime, J.P. Seasonal variation in the seed banks of herbaceous species in ten contrasting habitats. *J. Ecol.* **1979**, *67*, 893–921. [CrossRef]
85. Roberts, H.A. Seed banks in soils. *Adv. Appl. Biol.* **1981**, *6*, 1–55.
86. Thompson, K.; Bakker, J.P.; Bekker, R.M. *The Soil Seed Banks of North West Europe: Methodology, Density and Longevity*; Cambridge University Press: Cambridge, UK, 1997.
87. Bakker, J.P.; Poschlod, P.; Strykstra, R.J.; Bekker, R.M.; Thompson, K. Seed banks and seed dispersal: Important topics in restoration ecology. *Acta Bot. Neerl.* **1996**, *45*, 461–490. [CrossRef]
88. Williams, P.A. Ecology of the endangered herb *Scutellaria* novae-zelandiae. *N. Z. J. Ecol.* **1992**, *16*, 127–135.
89. Walck, J.L.; Baskin, J.M.; Baskin, C.C.; Hidayati, S.N. Defining transient and persistent seed banks. *Seed Sci. Res.* **2005**, *15*, 189–196. [CrossRef]
90. Finch-Savage, W.E.; Leubner-Metzger, G. Seed dormancy and the control of germination. *New Phytol.* **2006**, *171*, 501–523. [CrossRef]
91. Lamichhane, J.R.; Dürr, C.; Schwanck, A.A.; Robin, M.H.; Sarthou, J.P.; Cellier, V.; Messean, A.; Aubertot, J.N.; Debaeke, P. Integrated management of damping-off diseases. A Review. *Agron. Sustain. Dev.* **2017**, *37*, 10. [CrossRef]
92. Slot, M.; Poorter, L. Diversity of tropical tree seedling responses to drought. *Biotropica* **2007**, *39*, 683–690. [CrossRef]

horticulturae

MDPI

Article

Effects of Exogenous IBA on Cutting Propagation in Different Cutting Types of *Bougainvillea spectabilis* 'Yunnan Purple'

Die Hu [1] and Longsheng Hu [2,*]

[1] College of Horticulture & Gardening, Yangtze University, Jingzhou 434000, China; 501055@yangtzeu.edu.cn
[2] State Key Laboratory of Tree Genetics and Breeding, Chinese Academy of Forestry and Northeast Forestry University, Beijing 100091, China
* Correspondence: hu-ls@foxmail.com

Abstract

Bougainvillea spectabilis 'Yunnan Purple' is a relatively cold-tolerant variety, but its low pollen viability and poor seed set have limited large-scale reproduction. To establish an efficient propagation protocol, cuttings from three types of *Bougainvillea spectabilis* 'Yunnan Purple' were treated with varying concentrations of exogenous indole-3-butyric acid (IBA). Rooting parameters, growth indicators, and physiological metrics were measured, and the optimal treatment was identified through comprehensive membership function evaluation. The results showed that cutting types significantly influenced rooting, root development, plant growth, organic compound content (soluble sugars, starch, and protein), and abscisic acid (ABA) content. Conversely, IBA concentration significantly affected rooting, root architecture, polyphenol oxidase (PPO) activity, and the levels of indole-3-acetic acid (IAA) and Brassinolide (BR). This comprehensive evaluation identified lignified shoots (LS) treated with 100 mg/L IBA (LS-100) as the optimal protocol, which achieved a rooting rate of 63% and significantly improved root formation, plant growth, root activity, organic compound content, PPO activity, and the levels of IAA and BR. This study provides valuable insights and technical guidance for the large-scale cutting propagation of *Bougainvillea spectabilis* 'Yunnan Purple'.

Keywords: *Bougainvillea spectabilis* 'Yunnan Purple'; IBA; cutting type; cutting propagation; comprehensive membership function evaluation

1. Introduction

Bougainvillea spectabilis Willd., an evergreen woody climbing shrub of the Nyctaginaceae family, is commonly cultivated in warm, humid, and sunny environments [1]. It is widely used in landscaping due to its diverse cultivars, colorful bracts, and prolonged flowering period [2]. However, *Bougainvillea* species are generally sensitive to cold [3]. This limits their ability to overwinter outdoors in central and northern China [4]. In a previous study [5], *Bougainvillea spectabilis* 'Yunnan Purple' was identified as the most cold-tolerant cultivar in Hubei Province.

However, *Bougainvillea* propagation is constrained by narrow floral tubes, low pollen viability, and poor seed set, which hampers the mass production of high-quality seedlings [6]. To address these challenges, cutting propagation has been employed as an effective asexual method that relies on plant cell totipotency [7], preserving genetic integrity while maintaining desirable parental traits and enhancing multiplication efficiency [8].

The adventitious rooting capacity of cuttings is influenced by both intrinsic factors (such as cutting diameter, cutting strength, nutrient content, and hormone levels of cuttings) and external factors (including humidity, temperature, and exogenous hormone application) [9,10]. In general, one-year-old cuttings exhibit better rooting capacity than two-year-old cuttings due to their advantageous physiological conditions, including higher levels of endogenous hormones, enzyme activity, and nutrient content [11]. During cutting propagation, endogenous hormones, particularly Indole-3-acetic acid (IAA), abscisic acid (ABA), gibberellin (GA), brassinolide (BR), and zeatin riboside (ZR), play essential regulatory roles in adventitious root initiation, root development, and plant growth [12]. Moreover, oxidase enzymes, including peroxidase (POD), polyphenol oxidase (PPO), and superoxide dismutase (SOD), also contribute to root development. PPO enzymes promote callus and adventitious root formation by catalyzing the complexation of IAA with phenolic compounds [13], while SOD enzymes protect plant cells from oxidative damage by scavenging reactive oxygen species (ROS) during cutting propagation [14]. And POD enzymes stimulate the formation of root primordium cells and eliminate H2O2 content [15].

Furthermore, cuttings lose direct nutrient supply after being detached from the mother plant. Therefore, carbon and nitrogen compounds are essential for cuttings to serve as substantial energy and nutrient resources for callus formation and root development [16]. Some studies [17,18] have demonstrated that older cuttings exhibit higher survival rates than younger ones, attributed to their higher nutrient reserves.

Among external factors, hormonal regulation is the most effective approach to promote adventitious root formation [19]. IBA (indole-3-butyric acid), an exogenous hormone, is a natural precursor of IAA with high light stability, resulting in more stable and efficient adventitious root induction upon application [19]. The efficacy of IBA led to its widespread adoption in woody plant propagation, as optimal concentrations showed species-specific variation [11,20].

To identify optimal rooting treatments that integrate both intrinsic characteristics and external factors, researchers have increasingly employed the method combining principal component analysis with comprehensive membership function evaluation [21,22]. This approach enables the establishment of highly efficient cutting propagation systems. Therefore, the integrated approach was employed to assess the rooting capability of *B. spectabilis* 'Yunnan Purple' cuttings by comparing rooting capacity, growth traits, and physiological indicators across three cutting types under different IBA concentrations. The findings provide a scientific experimental methodology for rapid *B. spectabilis* 'Yunnan Purple' propagation, with both theoretical relevance and practical value.

2. Materials and Methods

2.1. Description of the Study Site

This experiment was conducted from April to August 2021 in a greenhouse at the West Campus of Yangtze University in Jingzhou, Hubei, China (112°15′ E, 30°36′ N). The greenhouse was maintained under a natural photoperiod (12/12 h, day/night), with temperature ranging from 20 °C (night minimum) to 41 °C (day maximum) and relative humidity between 70% and 85%.

2.2. Preparation of Experimental Materials

Healthy cuttings of *B. spectabilis* 'Yunnan Purple' were collected from five-year-old mother plants, sourced from Yibin Sijin Landscape Co., Ltd. in Yibin City, Sichuan Province, China.

Three types of cuttings were prepared (Figure 1): lignified shoots (LS; 13.2 ± 0.5 mm Ø), semi-lignified shoots (SS; 11.2 ± 0.5 mm Ø), and young lignified shoots (YS; 7.3 ± 0.5 mm Ø).

All cuttings were trimmed to 20 cm in length, with leaves removed and three latent buds retained. The base was cut at a 45° angle, and the top was trimmed flat. The growth substrate was an equal-volume mixture of river sand, perlite, vermiculite, and coco peat (Jiangsu Peilei Substrate Technology Co., Ltd., Suzhou, China). Sterilization was performed using 0.1% carbendazim for six hours (Shanghai Haohong Biomedicine Technology Co., Ltd., Shanghai, China). After that, cuttings were inserted into the substrate to one-third of their length, watered thoroughly by overhead sprinkling (the volume soil water content was approximately 37%), and the entire cutting bed was shaded with a shading net (50–70% nominal shade rate; Changzhou Green Netting Co., Ltd., Changzhou, China). Cuttings were watered via overhead sprinkling at 20 mL per plant every two days (after 48 h, the volume of soil water content was around 30%) under greenhouse conditions. All cuttings were maintained under greenhouse conditions for growth.

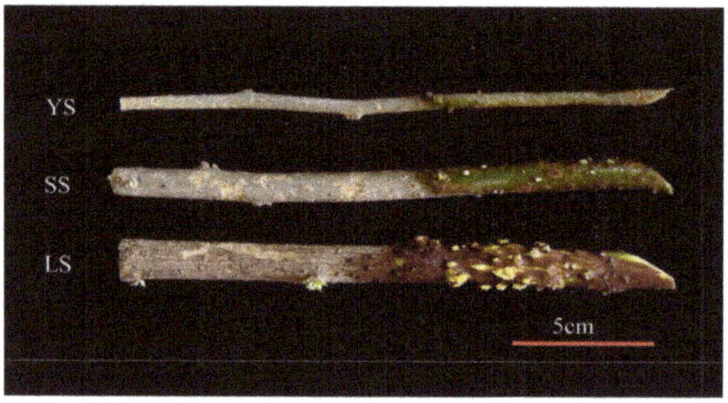

Figure 1. Lignified shoots (LS), Semi-lignified shoots (SS), and Young lignified shoots (YS) of *B. spectabilis* 'Yunnan Purple'.

2.3. Treatments and Research Design

The experiment employed a two-factor completely randomized design, incorporating four concentrations of exogenous IBA (CAS: 133-32-4, Merck KGaA, Darmstadt, Germany). IBA was initially dissolved in ethanol before being diluted with water. The design also included three types of cuttings, resulting in 12 treatment combinations (Table 1). Each treatment consisted of 40 potted cuttings. The experiment commenced on April 10, with each cutting receiving the basal drench of 20 mL of the corresponding IBA solution monthly, for a total of four applications over the experimental period. After 120 days, survival rate, rooting rate, and growth indicators were recorded. Additionally, leaf and root samples were immediately collected in liquid nitrogen for subsequent analysis of various physiological parameters.

Table 1. *B. spectabilis* 'Yunnan Purple' cuttings treatment.

Treatment	Lignified Shoots (LS)	Semi-Lignified Shoots (SS)	Young Lignified Shoots (YS)
IBA = 0 mg/L	LS-0	SS-0	YS-0
IBA = 50 mg/L	LS-50	SS-50	YS-50
IBA = 100 mg/L	LS-100	SS-100	YS-100
IBA = 150 mg/L	LS-150	SS-150	YS-150

2.4. Data Collection

The volume soil water content was monitored using a Spectrum TDR350 m (Spectrum Technologies Inc., New Orleans, LA, USA). Plants height and branch length of the cuttings were measured in centimeters using a straightedge. Rooting rate was evaluated at day 120, followed by root sample collection and root system analysis. Root morphological parameters, including total root length, total projected area, total surface area, average diameter, and total volume, were determined and analyzed using a WinRHIZO root analysis system (WinRHIZO version 2012b, Regent Instruments, Montreal, QC, Canada). Rooting percentage and rooting index were calculated as follows:

$$\text{Rooting percentage} = \text{rooted cuttings}/\text{total cuttings} \times 100 \quad (1)$$

$$\text{Rooting Index} = \text{rooting rate} \times \text{average root length} \times \text{average root number} \quad (2)$$

Additionally, the content of key organic compounds (Soluble sugar, soluble starch, and soluble protein content), along with superoxide dismutase (SOD) activity, peroxidase (POD) activity, and polyphenol oxidase (PPO) activity of leaves and the root activity were measured according to the methods described in the *Plant Physiology Experiment Tutorial* by Tang Shaohu [23]. Adventitious root samples from each treatment were collected, rinsed clean, and preserved in dry ice and sent to China Agricultural University for the determination of endogenous hormone levels using enzyme-linked immunosorbent assay (ELISA).

2.5. Data Processing and Analysis

The data were entered into Excel 2016 and analyzed using one-way analysis of variance (one-way ANOVA), Duncan's multiple comparison test, and principal component analysis (PCA) with IBM SPSS Statistics 27.0.1. Figures were prepared using Origin 2021.

3. Results

3.1. Rooting Analysis of B. spectabilis 'Yunnan Purple' Cuttings

3.1.1. Root System Morphology of *B. spectabilis* 'Yunnan Purple' Cuttings

Root growth morphology of *B. spectabilis* 'Yunnan Purple' cuttings was significantly influenced by IBA concentration and cutting type (Figure 2). Sparse root systems were observed in YS-type cuttings, compared to SS and LS types. SS-type cuttings produced fine and densely branched roots, whereas the LS types developed harder, longer, and more extensive root systems. Moreover, IBA treatment significantly improved root development and exhibited an extensive root system, particularly at concentrations of 100 and 150 mg/L, compared to the untreated control.

3.1.2. Rooting Indicators of *B. spectabilis* 'Yunnan Purple' Cuttings

In this study, a survival rate of 81% and an overall rooting rate of 42% were recorded for *B. spectabilis* 'Yunnan Purple' cuttings. Among the treatments, the maximum number of adventitious roots (30 and 27) was produced by SS-50 and SS-100, respectively, with no statistical differences among them; however, the minimum (10 and 13.33) was observed in YS-50 and YS-150, respectively, showing no significant difference (Table 2). The longest adventitious root length was observed with no statistical differences with all the results, including the letter 'a' in the same column (Table 2). The highest survival rate (98%) was achieved by LS-100, while no significant difference was shown among LS (0, 50,100, and 150) and SS-100 (Table 2). The maximum rooting rate (63%) was shared by LS-50 and LS-100, while the minimum (20%) was recorded in SS-0 and YS-0 (Table 2). The rooting index ranged from 18.44 (YS-0) to 126.47 (LS-100), the maximum rooting index was LS-100 (126.47) and SS-100 (111.93) (Table 2). Root fresh weight reached its maximum in LS-100 (3.57 g) and its minimum in YS-0 (0.68 g) and YS-50 (0.7 g) (Table 2). The maximum root length did not differ significantly among most treatments. Additionally, SS-0 differed significantly from both SS-50 and YS-100. (Table 2).

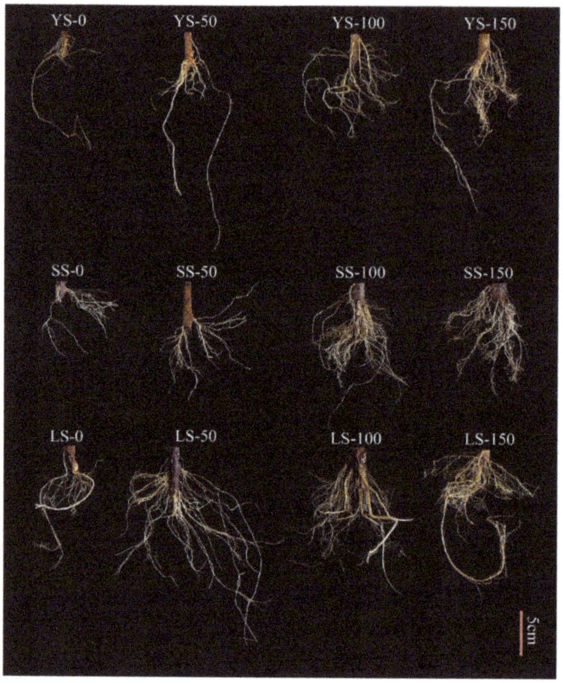

Figure 2. Root system morphology of *B. spectabilis* 'Yunnan Purple' under IBA treatment (0 mg/L, 50 mg/L, 100 mg/L, 150 mg/L) in Young lignified (YS), semi-lignified (SS), and lignified (LS) cuttings after 120 days.

Table 2. Rooting indexes of *B. spectabilis* 'Yunnan Purple' under IBA treatment (0 mg/L, 50 mg/L, 100 mg/L, 150 mg/L) in Young lignified (YS), semi-lignified (SS), and lignified (LS) cuttings after 120 days.

Treatment	Adventitious Root Numbers	Adventitious Root Length (cm)	Survival Rate (%)	Rooting Rate (%)	Rooting Index	Root Fresh Weight (g)	Maximum Root Length (cm)
YS-0	15.00 ± 1.00 d	4.73 ± 1.46 d	0.68 ± 0.07 bc	0.25 ± 0.05 de	18.44 ± 8.70 f	0.68 ± 0.02 i	23.50 ± 2.18 ab
YS-50	10.00 ± 0 e	6.43 ± 0.06 cd	0.65 ± 0.05 c	0.30 ± 0.05 d	19.30 ± 3.20 f	0.70 ± 0.01 i	24.00 ± 1.00 ab
YS-100	15.67 ± 3.06 cd	7.45 ± 1.39 abcd	0.73 ± 0.05 bc	0.40 ± 0.05 c	45.59 ± 6.20 def	1.58 ± 0.033 f	26.33 ± 9.29 a
YS-150	13.33 ± 3.79 de	9.68 ± 3.50 ab	0.78 ± 0.05 b	0.43 ± 0.03 bc	52.43 ± 7.87 de	1.48 ± 0.01 f	23.00 ± 6.08 ab
SS-0	17.67 ± 1.53 cd	6.20 ± 0.35 cd	0.68 ± 0.05 bc	0.20 ± 0.05 e	21.84 ± 5.83 f	1.15 ± 0.01 g	14.83 ± 1.61 b
SS-50	30.00 ± 1.00 a	8.50 ± 1.15 abc	0.76 ± 0.03 b	0.38 ± 0.03 c	98.39 ± 20.31 b	2.14 ± 0.08 c	26.00 ± 3.61 a
SS-100	27.00 ± 3.61 ab	8.33 ± 1.60 abc	0.93 ± 0.06 a	0.50 ± 0.05 b	111.93 ± 25.96 ab	2.79 ± 0.11 b	22.33 ± 7.23 ab
SS-150	24.67 ± 4.73 b	6.70 ± 0.70 bcd	0.77 ± 0.03 b	0.43 ± 0.03 bc	71.69 ± 15.83 cd	1.74 ± 0.07 e	19.67 ± 7.57 ab
LS-0	15.33 ± 2.52 cd	7.23 ± 1.15 abcd	0.95 ± 0.09 a	0.40 ± 0.05 c	43.38 ± 1.09 ef	0.97 ± 0.02 h	20.03 ± 2.55 ab
LS-50	17.33 ± 1.53 cd	8.57 ± 2.06 abc	0.88 ± 0.08 a	0.63 ± 0.06 a	94.19 ± 25.96 bc	1.90 ± 0.12 d	22.63 ± 2.14 ab
LS-100	20.00 ± 1.73 c	10.00 ± 1.22 a	0.98 ± 0.03 a	0.63 ± 0.03 a	126.47 ± 18.17 a	3.58 ± 0.17 a	23.77 ± 2.93 ab
LS-150	14.67 ± 1.53 d	8.33 ± 1.08 abc	0.92 ± 0.03 a	0.48 ± 0.03 b	59.16 ± 11.24 de	2.19 ± 0.15 c	17.87 ± 5.07 ab

Note: The above statistical data are the average of three replicates. Different lowercase letters in the same column indicate significant differences ($p < 0.05$).

3.1.3. Root Morphology Indexes of *B. spectabilis* 'Yunnan Purple' Cuttings

Significant differences were observed among different treatment groups of B. spectabilis 'Yunnan Purple' in terms of total root length, root projected area, total root surface area, total root volume, and average root diameter (Table 3). The combination of lignified shoots and 100 mg/L IBA (LS-100) was identified as the optimal condition for root system development. This treatment combination resulted in the maximum values for root projected area (40.83 cm^2), total root surface area (123.06 cm^2), and total root volume (5.35 cm^3). Total root length was maximized under SS-100 (532.33 mm) and LS-100 (497.28 mm), while the poorest performance was consistently observed across all parameters in the YS-0 control group.

Table 3. Root morphology indexes of cuttings of *B. spectabilis* 'Yunnan Purple' under IBA treatment (0 mg/L, 50 mg/L, 100 mg/L, 150 mg/L) in Young lignified (YS), semi-lignified (SS), and lignified (LS) cuttings after 120 days.

Treatment	Total Root Length (mm)	Root Projected Area (cm^2)	Total Root Surface Area (cm^2)	Total Root Volume (cm^3)	Average Root Diameter (mm)
YS-0	93.87 ± 19.30 f	7.07 ± 2.47 e	19.85 ± 5.83 e	0.55 ± 0.26 d	0.67 ± 0.12 cd
YS-50	112.13 ± 22.00 ef	8.72 ± 4.84 e	25.13 ± 11.67 e	0.70 ± 0.52 d	0.60 ± 0.13 d
YS-100	235.16 ± 73.30 cd	14.69 ± 5.72 de	46.17 ± 17.99 de	1.14 ± 0.63 cd	0.62 ± 0.14 cd
YS-150	197.25 ± 54.87 de	12.31 ± 5.47 de	38.67 ± 17.21 de	0.97 ± 0.68 d	0.62 ± 0.16 cd
SS-0	208.55 ± 40.77 de	12.41 ± 2.63 de	38.99 ± 8.29 de	1.01 ± 0.09 d	0.71 ± 0.03 bcd
SS-50	395.94 ± 57.93 b	31.04 ± 5.64 b	97.06 ± 18.36 ab	2.87 ± 0.98 b	0.83 ± 0.09 abc
SS-100	532.33 ± 48.57 a	32.6 ± 6.43 ab	102.42 ± 20.22 ab	2.46 ± 0.96 bc	0.62 ± 0.05 cd
SS-150	330.31 ± 45.96 bc	20.78 ± 3.56 cd	65.29 ± 11.20 cd	1.74 ± 0.45 bcd	0.69 ± 0.02 cd
LS-0	245.43 ± 69.88 cd	15.39 ± 2.00 de	48.36 ± 6.28 de	1.22 ± 0.38 cd	0.68 ± 0.13 cd
LS-50	391.32 ± 59.43 b	25.46 ± 5.82 bc	80.18 ± 18.06 bc	2.67 ± 1.45 b	0.73 ± 0.06 abcd
LS-100	497.28 ± 66.05 a	40.83 ± 4.50 a	123.06 ± 23.03 a	5.35 ± 0.19 a	0.90 ± 0.03 ab
LS-150	351.21 ± 69.62 b	32.3 ± 6.88 ab	102.06 ± 20.70 a	4.58 ± 0.90 a	0.93 ± 0.18 a

Note: The above statistical data are the average of three replicates. Different lowercase letters in the same column indicate significant differences ($p < 0.05$).

3.2. Growth Analysis of B. spectabilis 'Yunnan Purple' Cuttings

3.2.1. Growth Morphology of *B. spectabilis* 'Yunnan Purple' Cuttings

The growth morphology of *B. spectabilis* 'Yunnan Purple' cuttings was significantly influenced by both exogenous IBA concentration and cutting type (Figure 3). Across all cutting types, IBA-treated cuttings exhibited superior growth, with more and longer branches compared to the untreated controls. Furthermore, cuttings derived from LS and SS types demonstrated significantly better growth performance than those from YS branches. These YS cuttings produced fewer and shorter branches.

Figure 3. Growth morphology of *B. spectabilis* 'Yunnan Purple' under IBA treatment (0, 50, 100, 150 mg/L) after 120 days. (**a**) Young lignified (YS), (**b**) semi-lignified (SS), and (**c**) lignified (LS) Cuttings.

3.2.2. Growth Indicators of *B. spectabilis* 'Yunnan Purple' Cuttings

The maximum plant height was recorded in LS-150 and SS-100, significantly differing from the minimum observed in YS-0 and in others as well. The highest plant in YS, SS, and LS cuttings was YS-100, SS-100, and LS-150, respectively (Table 4). Branch numbers varied significantly with cutting type. Specifically, LS cuttings consistently produced more branches than SS or YS types. Furthermore, no significant differences in branch numbers were associated with IBA concentration. The maximum branch numbers were recorded in LS-0, LS-100, and LS-150, with no significant differences among these treatments (Table 4). Branch length, however, was affected by both cutting type and IBA concentration. The longest branches were observed in SS-100 and LS-50, which were significantly longer than those in YS-150, recording the shortest branches (Table 4).

Table 4. Growth indicators of Cuttings of *B. spectabilis* 'Yunnan Purple' under IBA treatment (0, 50, 100, 150 mg/L) in Young lignified (YS), semi-lignified (SS), and lignified (LS) Cuttings after 120 days.

Treatment	Plant Height (cm)	Branch Numbers	Branch Length (cm)
YS-0	21.68 ± 2.95 d	2.19 ± 0.99 c	5.95 ± 3.39 cd
YS-50	21.73 ± 3.04 d	2.50 ± 0.99 bc	6.56 ± 4.82 bcd
YS-100	22.45 ± 3.76 cd	2.48 ± 0.91 bc	5.70 ± 3.58 cd
YS-150	21.82 ± 2.28 d	2.19 ± 0.91 c	5.34 ± 3.04 d
SS-0	23.61 ± 3.39 cd	2.04 ± 0.77 c	9.02 ± 5.42 abc
SS-50	24.43 ± 4.56 bc	2.26 ± 1.26 c	9.76 ± 6.09 ab
SS-100	27.62 ± 4.61 a	2.36 ± 1.06 bc	11.95 ± 7.88 a
SS-150	24.60 ± 5.19 bc	2.44 ± 1.16 bc	8.61 ± 5.34 bcd
LS-0	24.98 ± 5.05 bc	3.28 ± 1.45 a	7.33 ± 5.51 bcd
LS-50	23.92 ± 4.39 cd	3.00 ± 1.06 ab	11.83 ± 9.17 a
LS-100	26.56 ± 5.14 ab	3.26 ± 1.36 a	8.99 ± 5.45 abc
LS-150	28.60 ± 6.71 a	3.22 ± 1.42 a	7.44 ± 5.33 bcd

Note: The above statistical data are the average of three replicates. Different lowercase letters in the same column indicate significant differences ($p < 0.05$).

3.3. *Physiological and Biochemical Analysis in B. spectabilis 'Yunnan Purple' Cuttings*

3.3.1. Root Activity in *B. spectabilis* 'Yunnan Purple'

The apical 1 cm segment of the root tips (Figure 4a) from *B. spectabilis* 'Yunnan Purple' cuttings was used for the measurement of root activity. It was observed (Figure 4b) that root activity was higher in LS and SS than in YS cuttings. Compared to the control groups without IBA, root activity significantly increased in most IBA-treated cuttings (YS, SS, and LS), except for the YS-50. The optimal IBA concentration for root activity was 150 mg/L in YS, 50 and 100 mg/L in SS, and 100 mg/L in LS.

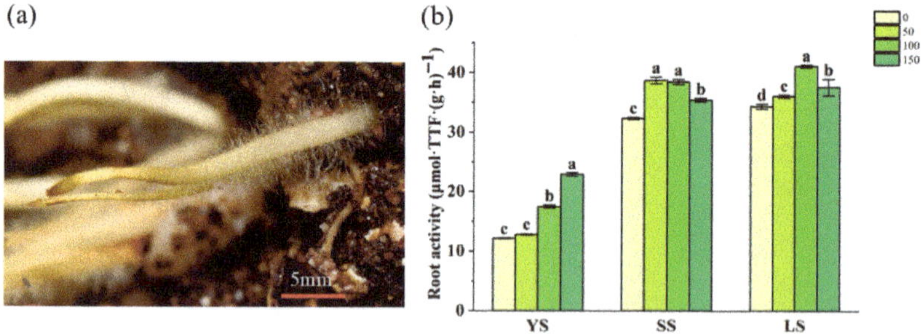

Figure 4. Observations of root tip morphology (**a**) and analysis of root system activity (**b**) in *B. spectabilis* 'Yunnan Purple' under IBA treatment (0, 50, 100, 150 mg/L) in Young lignified (YS), semi-lignified (SS), and lignified (LS) Cuttings after 120 days. Note: The above statistical data are the average of three replicates. Different lowercase letters in the same column indicate significant differences ($p < 0.05$).

3.3.2. Oxidase Activity in *B. spectabilis* 'Yunnan Purple'

The oxidase activities in *B. spectabilis* 'Yunnan Purple' were measured (Figure 5).

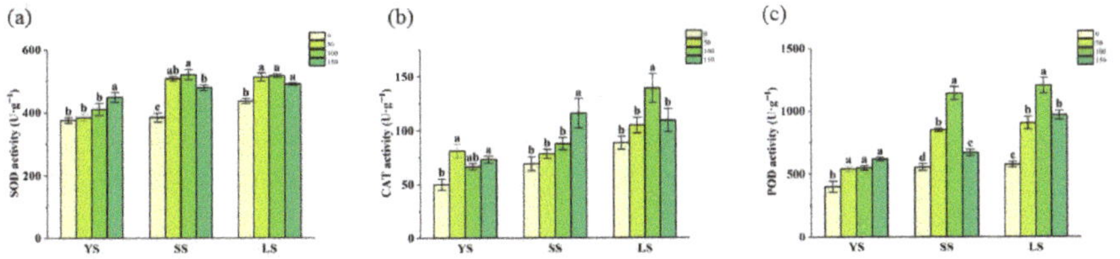

Figure 5. SOD (**a**), CAT (**b**), and POD (**c**) activities in *B. spectabilis* 'Yunnan Purple' under IBA treatment (0, 50, 100, 150 mg/L) in Young lignified (YS), semi-lignified (SS), and lignified (LS) Cuttings after 120 days. Note: The above statistical data are the average of three replicates. Different lowercase letters in the same column indicate significant differences ($p < 0.05$).

The highest SOD activity (Figure 5a) was observed in YS at 150 mg/L IBA and in SS at 50 and 100 mg/L IBA, respectively. However, in LS cuttings, SOD activity showed no significant response to IBA treatments, while it was markedly higher than in the no-hormone treatment.

For the CAT activity, the optimal IBA concentrations were 150 mg/L for SS and 100 mg/L for LS. Nevertheless, no significant difference was observed among the 50, 100, and 150 mg/L IBA treatments for YS.

In YS cuttings, POD activity did not differ among the 50, 100, and 150 mg/L IBA treatments. However, all IBA-treated groups exhibited significantly higher activity than the control. In contrast, both SS and LS cuttings reached their maximum POD activity at 100 mg/L IBA.

3.3.3. Organic Compound Content in *B. spectabilis* 'Yunnan Purple'

Organic compound content analysis of *B. spectabilis* 'Yunnan Purple' cuttings indicated that SS and LS exhibited elevated levels of soluble proteins (Figure 6b) and starch (Figure 6c) compared to YS, with the untreated groups showing the lowest values. The highest soluble sugar (Figure 6a) content was observed in YS with 100 and 150 mg/L IBA, in SS with 50 and 100 mg/L IBA, and in LS with 100 mg/L IBA. The optimal IBA concentrations for soluble protein (Figure 6b) content peaked at 150 mg/L in YS and at 100 mg/L in both SS and LS. Similarly, the soluble starch (Figure 6c) content reached its highest level at 100 and 150 mg/L IBA in YS, 50 and 100 mg/L IBA in SS, and at 100 mg/L in LS.

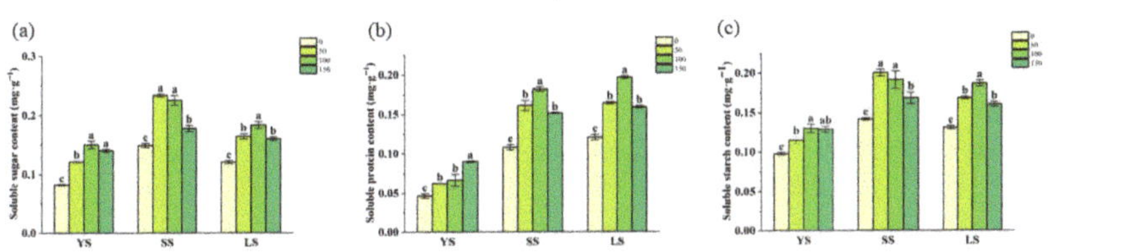

Figure 6. Soluble sugar (**a**), soluble protein (**b**), and soluble starch (**c**) content in *B. spectabilis* 'Yunnan Purple' under IBA treatment (0, 50, 100, 150 mg/L) in Young lignified (YS), semi-lignified (SS), and lignified (LS) Cuttings after 120 days. Note: The above statistical data are the average of three replicates. Different lowercase letters in the same column indicate significant differences ($p < 0.05$).

3.3.4. Endogenous Hormone Content in *B. spectabilis* 'Yunnan Purple'

Relative to the control, the application of IBA significantly increased the contents of Indole-3-acetic acid (IAA, Figure 7a) and Gibberellin A3 (GA3, Figure 7b) in YS, SS, and LS cuttings. Specifically, the IAA content peaked at 100 mg/L IBA in LS cuttings, showing a significant advantage over other treatments. The GA3 content reached its maximum at 100 and 150 mg/L IBA in YS cuttings, and at 50 and 100 mg/L IBA in both SS and LS cuttings. Similarly, the Zeatin riboside (ZR) content (Figure 7c) peaked at 150 mg/L IBA in YS cuttings, and at 50 and 100 mg/L IBA in SS and LS cuttings. The Brassinolide (BR) content (Figure 7d) exhibited peaks at 100 and 150 mg/L IBA in YS and LS cuttings, while in SS cuttings, the peak was observed specifically at 150 mg/L IBA. In contrast, abscisic acid (ABA) content (Figure 7e) presented a distinct pattern. Its relatively highest levels were found at 0 and 50 mg/L IBA in YS cuttings, and at 0 and 150 mg/L IBA in both SS and LS cuttings.

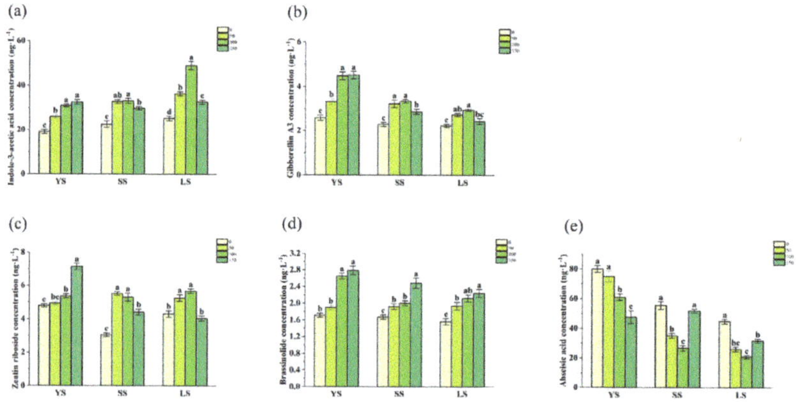

Figure 7. Indole-3-acetic acid (**a**), Gibberellin A3 (**b**), Zeatin riboside (**c**), Brassinolide (**d**), and Abscisic acid (**e**) concentration in *B. spectabilis* 'Yunnan Purple' under IBA treatment (0, 50, 100, 150 mg/L) in Young lignified (YS), semi-lignified (SS), and lignified (LS) Cuttings after 120 days. Note: The above statistical data are the average of three replicates. Different lowercase letters in the same column indicate significant differences ($p < 0.05$).

3.4. Analysis of Variance and Comprehensive Membership Function Evaluation

3.4.1. Variance Analysis of *B. spectabilis* 'Yunnan Purple'

Variance analysis was conducted on 26 indices. The results (Table S1) revealed that the number of adventitious roots, survival rate, rooting percentage, rooting index, total root length, root projected area, total root surface area, total root volume, and root fresh weight were all significantly positively affected by both the cutting type and exogenous hormone concentration. In contrast, adventitious root length, PPO activity, IAA concentration, and BR concentration showed significant positive correlation with exogenous hormone levels. Meanwhile, average root diameter, plant height, branch length, soluble sugar content, soluble starch content, and soluble protein content were significantly positively associated with cutting types. Conversely, ABA concentration exhibited a significantly negative association.

3.4.2. Principal Component Analysis and Comprehensive Evaluation of *B. spectabilis* 'Yunnan Purple'

Principal component analysis (PCA) was performed on 26 indices related to rooting, growth, and physiological characteristics of *B. spectabilis* 'Yunnan Purple' cuttings (Table 5). The first five principal components, which together accounted for 94.51% of the total variance, were selected for comprehensive membership function evaluation. The results indicated that the LS-100 treatment scored the highest, followed by SS-100. In contrast, the YS-0 and YS-50 showed the lowest scores.

Table 5. Comprehensive evaluation of rooting using membership function for *B. spectabilis* 'Yunnan Purple' in Young lignified (YS), semi-lignified (SS), and lignified (LS) Cuttings under IBA treatment (0, 50, 100, 150 mg/L).

Treatment	Principal Components 1	Principal Components 2	Principal Components 3	Principal Components 4	Principal Components 5	Score
LS-100	6.895	0.4696	−2.0532	0.4983	0.8912	0.8008
SS-100	4.9761	0.1486	2.5079	−1.1337	−0.5464	0.7022
SS-50	3.0817	0.3584	2.4971	1.2107	1.4129	0.6152
LS-50	3.4504	−0.5321	−0.4006	−2.0484	−0.0873	0.6120
LS-150	2.4735	−1.7405	−2.5673	1.7629	−0.4829	0.5720
SS-150	0.4109	−0.349	1.4637	0.5917	−0.8565	0.4771
YS-100	−2.4018	3.0904	0.0801	0.2010	0.2110	0.3656
YS-150	−2.0869	4.1003	−0.9382	0.1181	−1.4684	0.3609
LS-0	−1.6002	−2.1371	−1.546	−1.6635	−0.0006	0.3403
SS-0	−3.6431	−3.1948	1.2381	0.8803	−1.262	0.2152
YS-50	−5.1249	0.5731	−0.327	−0.3610	0.7564	0.1894
YS-0	−6.4308	−0.7868	0.0453	−0.0564	1.4325	0.1250
Contribution	62.04	14.71	9.79	4.61	3.36	/

4. Discussion

4.1. Cutting Type as a Key Intrinsic Factor Affecting the Rooting and Growth of B. spectabilis *'Yunnan Purple'*

The regenerative and rooting potential of cuttings is positively influenced by the accumulation of organic compounds and endogenous hormones [24]. Generally, in highly lignified cuttings, metabolic processes are commonly weakened, and rooting capacity is significantly reduced by the combined effects of accumulating inhibitory compounds (such as tannins and phenolics) and declining auxin levels [25].

In our study, lignified (LS) and semi-lignified (SS) shoot cuttings exhibited superior rooting performance, showing higher rooting rates, increased root activity, and more developed root systems than young semi-lignified shoot cuttings (YS). Variance analysis confirmed that cutting type significantly influenced the rooting rate and root growth, consistent with previous reports of a positive correlation between cutting diameter and rooting rate [26]. Meanwhile, plant height and branch length were significantly correlated with cutting type. Specifically, LS cuttings yielded the tallest plants with the most branches, outperforming both SS and YS. Branch length was also greater in both LS and SS cuttings compared to YS. These findings demonstrate that cutting age and diameter collectively enhance plant growth, as older and thicker cuttings provide more nutrient reserves that support superior branch development [27,28].

Furthermore, adventitious root formation in plant cuttings is an energy-intensive process that demands substantial material resources. Carbon and nitrogen compounds are essential substrates during rooting, providing critical nutritional support for callus formation and subsequent root development [29,30]. In this study, soluble sugars, proteins, and starch showed significant correlations with cutting type, with SS and LS cuttings containing higher levels of these nutrients than YS cuttings. Soluble starch serves as a hydrolyzable source of soluble sugars [31], while the sugars themselves provide both energy and signaling functions that stimulate adventitious rooting and growth [32]. Soluble proteins act as structural components of protoplasm while also functioning as enzymatic regulators involved in morphogenesis, cell growth, and differentiation [33]. Meanwhile, some studies suggest that carbohydrate content may not directly control rooting rate but rather regulates root growth and development [34,35].

4.2. Exogenous IBA Significantly Enhances the Rooting and Growth of Cuttings by Regulating Enzymatic Activity and Endogenous Hormone Homeostasis

Exogenous IBA plays a key regulator of adventitious rooting by converting to IAA, thereby enhancing auxin biosynthesis and transport [36]. This regulation involves synergistic actions of peroxisome-derived metabolites and jasmonates, along with downstream auxin signaling components [36].

Our study demonstrated that most IBA-treated cuttings exhibited markedly higher rooting rates, enhanced root activity, and superior root development compared to the control group. These findings align with practical observations that IBA application significantly shortens the rooting period, increases root number, improves root architecture, and enhances root vitality relative to untreated controls [37,38].

Studies on tea cuttings have confirmed that optimal IBA concentrations promote basal cell division, stimulate root primordium differentiation, and enhance vascular development [19]. In our experiments, superior rooting was achieved with 100 mg/L IBA compared to 150 mg/L in both SS and LS cuttings, confirming the importance of precise concentration optimization to avoid both deficiency and supra-optimal inhibition.

Meanwhile, cuttings undergo post-excision stress from interrupted nutrient supply, making it vital to enhance their stress tolerance and accelerate root emergence. Antioxidant enzymes such as POD, SOD, and PPO play crucial roles in oxidative defense and root development.

The root formation phase was characterized by enhanced POD activity, improving stress resistance and adventitious rooting, along with markedly increased IAA content [29]. Moreover, SOD scavenges reactive oxygen species to alleviate oxidative stress, delay senescence, and promote rooting [39]. Furthermore, upregulated PPO activity promotes rooting by protecting cuttings and facilitating root formation [12].

Our studies demonstrated that exogenous IBA significantly enhanced the activity of antioxidant enzymes. However, only PPO activity was significantly correlated with IBA concentration, suggesting that rooting promotion by IBA occurred primarily through PPO upregulation. These findings were consistent with reports in tea tree propagation that IBA treatment also significantly increased PPO activity, which improved rooting efficiency [40,41]. These results confirm the crucial role of antioxidant enzymes in cuttings' survival and root development.

Endogenous hormone balance also plays a central role in adventitious root formation. Although no significant correlation was observed between rooting-promoting hormone content and cutting type, significant correlations were found between IBA concentration and both IAA and BR levels. Both IAA and BR levels were increased by exogenous IBA application, indicating that rooting was primarily facilitated through IBA-mediated modulation of endogenous IAA and BR levels.

Adventitious root formation was primarily induced by IAA [42]. IAA initiates root formation by binding to cell surface receptors, triggering signaling pathways that activate genes responsible for cell division and differentiation [37]. This process leads to callus formation and establishes the cellular foundation for root primordium development [43]. Meanwhile, BR regulated the differentiation and division of root tip meristem cells. Low BR concentrations promoted rooting, while high concentrations produced inhibitory effects. A significant synergistic effect is observed between BRs and auxin during root development, potentially mediated through overlapping activities, shared target genes, or enhanced auxin transport [44].

4.3. LS-100 Treatment Showed Optimal Rooting in B. spectabilis 'Yunnan Purple' Cuttings by Membership Function Analysis

The principal component analysis (PCA) was used to reduce a large number of variables to a smaller group of variables that explained performance; the performance cues were used for further comprehensive evaluation [45]. The weight of each indicator was objectively determined through this method, effectively minimizing the influence of overlapping information [46]. The method has been widely used for rooting evaluation in various plant species [45,47]. In this study, the first five principal components accounted for 94.06% were selected for comprehensive membership function evaluation. The LS-100 treatment group (lignified cuttings treated with 100 mg/L IBA) exhibited the highest membership function value, indicating its optimal overall rooting performance.

Most *Bougainvillea* cultivars are inherently hard-to-root, limiting the effectiveness of cutting propagation despite its wide use in ornamental species [48]. Studies observed that IBA concentrations of 6000, 100, and 2000 ppm significantly increased the rooting performance of stem cuttings of *Bougainvillea* 'Mahara', 'Akola,' and 'Thimma', respectively [48,49]. Interestingly, the hardwood cuttings of *Bougainvillea* produced the highest number and the longest roots compared to semi-hardwood and softwood cuttings [48,50]. Those conclusions are consistent with our findings. However, the relatively low rooting rate in this study was likely derived from high temperatures in the greenhouse, which could have adversely affected root formation [51,52].

In summary, our findings provide technical guidance for *B. spectabilis* 'Yunnan Purple' propagation with substantial practical applications. A promising direction for future research would be the investigation of temperature effects on rooting efficiency.

5. Conclusions

This study demonstrates that the cutting propagation of *B. spectabilis* 'Yunnan Purple' from April to August is co-determined by cutting type and exogenous IBA concentration. The key parameters showed that adventitious root number, rooting rate, root system architecture (total length, projected area, volume), and root fresh weight differed significantly among cutting types and were further enhanced by IBA application relative to their respective controls. Plant growth (plant height, branch length) and the contents of soluble sugars, starch, and protein were predominantly determined by cutting type. In contrast, the activity of polyphenol oxidase (PPO) was significantly influenced only by IBA application. At the hormonal level, IBA treatment increased the endogenous levels of indole-3-acetic acid (IAA) and Brassinolide (BR), with cutting type concurrently influencing abscisic acid (ABA) content. According to the comprehensive membership function evaluation, the treatment of lignified shoots (LS) with 100 mg/L IBA (LS-100) resulted in optimal protocol, achieving a rooting rate of 63% and significantly improved root formation, plant growth, root activity, organic compound content (soluble sugars, starch, and protein), PPO activity, and the levels of IAA and BR. These findings provide theoretical support and technical guidance for the large-scale propagation of *B. spectabilis* 'Yunnan Purple'.

Supplementary Materials: The following supporting information can be downloaded at: https://www.mdpi.com/article/10.3390/horticulturae11121496/s1, Table S1: Variance analysis on rooting of *B. spectabilis* 'Yunnan Purple' under IBA treatment (0, 50, 100, 150 mg/L) in Young lignified (YS), semi-lignified (SS), and lignified (LS) Cuttings after 120 days.

Author Contributions: L.H.: Conceptualization, formal analysis, writing—original draft preparation, funding acquisition; D.H.: supervision and writing—review and editing. All authors have read and agreed to the published version of the manuscript.

Funding: This research was funded by the Science and Technology Development Program of Jingzhou, Hubei Province, A Key Technology for the Rapid Propagation of Cold-Tolerant Bougainvillea Cultivars, via grant numbers (2021CC28-24).

Data Availability Statement: The original contributions presented in this study are included in the article/Supplementary Material. Further inquiries can be directed to the corresponding author.

Acknowledgments: The authors are greatly indebted to the Science and Technology Development Program of Jingzhou, Hubei Province, for their financial support and the Department of Horticulture and Landscape Architecture college, University of Yangtze, for their technical support and help with experiment preparation.

Conflicts of Interest: The authors declare no conflicts of interest. The funders had no role in the design of the study; in the collection, analyses, or interpretation of data; in the writing of the manuscript; or in the decision to publish the results.

Abbreviations

The following abbreviations are used in this manuscript:

LS	Lignified shoots
SS	Semi-lignified shoots
YS	Young lignified shoots
ZR	Zeatin riboside
IAA	Indole-3-acetic acid
BR	Brassinolide
GA_3	Gibberellic acid
ABA	Abscisic acid
ROS	Scavenging reactive oxygen species
PPO	Polyphenol oxidase
POD	Peroxidase
SOD	Superoxide dismutase
PCA	Principal component analysis

References

1. Bortolotti, M.; Bolognesi, A.; Polito, L. Bouganin, an Attractive Weapon for Immunotoxins. *Toxins* **2018**, *10*, 323. [CrossRef]
2. Lin, H.; Xu, J.; Wu, K.; Gong, C.; Jie, Y.; Yang, B.; Chen, J. An Efficient Method for the Propagation of *Bougainvillea glabra* 'New River' (*Nyctaginaceae*) from in Vitro Stem Segments. *Forests* **2024**, *15*, 519. [CrossRef]
3. Zhang, H.; Huang, T.; Zhou, Q.; Sheng, Q.; Zhu, Z. Complete Chloroplast Genomes and Phylogenetic Relationships of *Bougainvillea spectabilis* and *Bougainvillea glabra (Nyctaginaceae)*. *Int. J. Mol. Sci.* **2023**, *24*, 13044. [CrossRef]
4. Ahmad, I.; Zamir, R.; Shah, S.T.; Wali, S. In Vitro Surface Sterilization of the Shoot Tips of *Bougainvillea spectabilis* Willd. *Pure Appl. Biol.* **2016**, *5*, 781–786. [CrossRef]
5. Li, M.X.; Wang, M.; Zhang, Y.; Li, L.L.; Hu, D. Effect of Spraying Cold Resistance Agents on Cold Resistance of *Bougainvillea glabra*. *Acta Agric. Jiangxi* **2020**, *32*, 55–60.
6. Yun, L.R.; Hua, Z.Z.; Qin, Z.X.; Xian, L.L.; Hua, S.M.; Nan, X.Z. Effects of Remaining Leaf Combining with IBA on Rooting, Physiological and Biochemical Indicators of Leaves from *Bougainvillea spectabilis* Cuttings. *Sci. Technol.* **2010**, *23*, 120–123.
7. Dong, Y.; Guo, W.; Xiao, W.; Liu, J.; Jia, Z.; Zhao, X.; Jiang, Z.; Chang, E. Effects of Different Donor Ages on the Growth of Cutting Seedlings Propagated from Ancient *Platycladus orientalis*. *Plants* **2023**, *12*, 1754. [CrossRef]
8. Eed, A.M.; Albana'a, B.; Almaqtari, S. The Effect of Growing Media and Stem Cutting Type on Rooting and Growth of Bougainvillea Spectabilis Plants. *Univ. Aden J. Nat. Appl. Sci.* **2015**, *19*, 141–147. [CrossRef]
9. Dob, A.; Lakehal, A.; Novak, O.; Bellini, C. Jasmonate Inhibits Adventitious Root Initiation through Repression of *CKX1* and Activation of *RAP2.6L* Transcription Factor in *Arabidopsis*. *J. Exp. Bot.* **2021**, *72*, 7107–7118. [CrossRef]
10. Huang, G.; Li, Y.; Zhang, Y.; Wen, W.; Zhao, C.; Guo, X. Overcoming Challenges in Plant Biomechanics: Methodological Innovations and Technological Integration. *Adv. Sci.* **2025**, *12*, 2415606. [CrossRef]
11. Wannasaksri, W.; Temviriyanukul, P.; Aursalung, A.; Sahasakul, Y.; Thangsiri, S.; Inthachat, W.; On-Nom, N.; Chupeerach, C.; Pruesapan, K.; Charoenkiatkul, S.; et al. Influence of Plant Origins and Seasonal Variations on Nutritive Values, Phenolics and Antioxidant Activities of Adenia Viridiflora Craib., an Endangered Species from Thailand. *Foods* **2021**, *10*, 2799. [CrossRef]

12. Zhou, X.; Li, R.; Shen, H.; Yang, L. Effect of Exogenous Plant Growth Regulators and Rejuvenation Measures on the Endogenous Hormone and Enzyme Activity Responses of Acer Mono Maxim in Cuttage Rooting. *Int. J. Mol. Sci.* **2023**, *24*, 11883. [CrossRef]
13. Somkuwar, R.G.; Bondage, D.D.; Surange, M.S.; Ramteke, S.D. Rooting Behaviour, Polyphenol Oxidase Activity, and Biochemical Changes in Grape Rootstocks at Different Growth Stages. *Turk. J. Agric. For.* **2011**, *35*, 281–287. [CrossRef]
14. Alscher, R.G.; Erturk, N.; Heath, L.S. Role of Superoxide Dismutases (SODs) in Controlling Oxidative Stress in Plants. *J. Exp. Bot.* **2002**, *53*, 1331–1341. [CrossRef]
15. Wang, H.; Xu, R.; You, L.; Zhong, G. Characterization of Cu-Tolerant Bacteria and Definition of Their Role in Promotion of Growth, Cu Accumulation and Reduction of Cu Toxicity in *Triticum aestivum* L. *Ecotoxicol. Environ. Saf.* **2013**, *94*, 1–7. [CrossRef]
16. Niu, L.; Yu, J.; Liao, W.; Xie, J.; Yu, J.; Lv, J.; Xiao, X.; Hu, L.; Wu, Y. Proteomic Investigation of S-Nitrosylated Proteins During NO-Induced Adventitious Rooting of Cucumber. *Int. J. Mol. Sci.* **2019**, *20*, 5363. [CrossRef]
17. Raviv, M.; Reuveni, O.; Goldschmidt, E.E. The Physiological Basis for Loss of Root ability with Age in Avocado Seedlings. *Tree Physiol.* **1987**, *3*, 115–122. [CrossRef] [PubMed]
18. Zhou, F.; Wang, N.; Zhang, J.; Yao, X.; Zhang, T.; Zhang, X.; Zhan, L.; Li, J. Formulation of Substrates with Agricultural and Forestry Wastes for Camellia Oleifera Abel Seedling Cultivation. *PLoS ONE* **2022**, *17*, e0265979. [CrossRef] [PubMed]
19. Damodaran, S.; Strader, L.C. Protocol for Determining Regulators of Competence in Regeneration of Adventitious Roots Using Hypocotyl Wounding Approach in *Arabidopsis*. *STAR Protoc.* **2025**, *6*, 103818. [CrossRef] [PubMed]
20. Singh, K.K.; Rawat, J.M.S.; Tomar, Y.K. Influence of IBA on Rooting Potential of Torch Glory *Bougainvillea*. *J. Hortic. Sci. Ornam. Plants* **2011**, *3*, 162–165.
21. Ding, C.; Xu, C.; Lu, B.; Zhu, X.; Luo, X.; He, B.; Elidio, C.; Liu, Z.; Ding, Y.; Yang, J.; et al. Comprehensive Evaluation of Rice Qualities under Different Nitrogen Levels in South China. *Foods* **2023**, *12*, 697. [CrossRef]
22. Yuan, J.; Li, Y.; Luo, Y.; Xuan, Y.; He, N. Evaluation of Cutting Growth and Rooting Potentials of Different Mulberry Varieties. *J. Southwest Univ. Nat. Sci. Ed.* **2025**, *47*, 61–69. [CrossRef]
23. Tang, S.; Luo, C. *Plant Physiology Experiment Tutorial*; Southwest China Normal University Press: Chongqing, China, 2012.
24. Liang, B.; Sun, Y.; Li, Z.; Zhang, X.; Yin, B.; Zhou, S.; Xu, J. Crop Load Influences Growth and Hormone Changes in the Roots of "Red Fuji" Apple. *Front. Plant Sci.* **2020**, *11*, 665. [CrossRef]
25. Di Battista, F.; Maccario, D.; Beruto, M.; Grauso, L.; Lanzotti, V.; Curir, P.; Monroy, F. Metabolic Changes Associated to the Unblocking of Adventitious Root Formation in Aged, Rooting-Recalcitrant Cuttings of *Eucalyptus gunnii* Hook. f. (*Myrtaceae*). *Plant Growth Regul.* **2019**, *89*, 73–82. [CrossRef]
26. Zhang, W.; Ma, Y.; Wang, Z.; Ye, X.; Shui, J. Some Soybean Cultivars Have Ability to Induce Germination of Sunflower Broomrape. *PLoS ONE* **2013**, *8*, e59715. [CrossRef]
27. Bannoud, F.; Bellini, C. Adventitious Rooting in Populus Species: Update and Perspectives. *Front. Plant Sci.* **2021**, *12*, 668837. [CrossRef]
28. Otiende, M.A.; Nyabundi, J.O.; Ngamau, K.; Opala, P. Effects of Cutting Position of Rose Rootstock Cultivars on Rooting and Its Relationship with Mineral Nutrient Content and Endogenous Carbohydrates. *Sci. Hortic.* **2017**, *225*, 204–212. [CrossRef]
29. Gao, J.; Zeng, X.; Liu, X.; Yang, S. Cutting propagation of *Periploca forrestii* and Dynamic Analyses of Physiological and Biochemical Characteristitics Related to Adventitious Roots Formation. *Zhong Yao Cai* **2011**, *34*, 841–845. [PubMed]
30. Opdahl, L.J.; Lewis, R.W.; Kalcsits, L.A.; Sullivan, T.S.; Sanguinet, K.A. Plant Uptake of Lactate-Bound Metals: A Sustainable Alternative to Metal Chlorides. *Biomolecules* **2021**, *11*, 1085. [CrossRef]
31. Chen, E.; Qin, L.; Li, F.; Yang, Y.; Liu, Z.; Wang, R.; Yu, X.; Niu, J.; Zhang, H.; Wang, H.; et al. Physiological and Transcriptomic Analysis Provides Insights into Low Nitrogen Stress in Foxtail Millet (*Setaria italica* L.). *Int. J. Mol. Sci.* **2023**, *24*, 16321. [CrossRef] [PubMed]
32. Monder, M.J.; Pacholczak, A. Rhizogenesis and Concentration of Carbohydrates in Cuttings Harvested at Different Phenological Stages of Once-Blooming Rose Shrubs and Treated with Rooting Stimulants. *Biol. Agric. Hortic.* **2020**, *36*, 53–70. [CrossRef]
33. Liu, Q.; Jones, C.S.; Parsons, A.J.; Xue, H.; Rasmussen, S. Does Gibberellin Biosynthesis Play a Critical Role in the Growth of *Lolium perenne*? Evidence from a Transcriptional Analysis of Gibberellin and Carbohydrate Metabolic Genes after Defoliation. *Front. Plant Sci.* **2015**, *6*, 944. [CrossRef] [PubMed]
34. Jan, M.; Muhammad, S.; Jin, W.; Zhong, W.; Zhang, S.; Lin, Y.; Zhou, Y.; Liu, J.; Liu, H.; Munir, R.; et al. Modulating Root System Architecture: Cross-Talk between Auxin and Phytohormones. *Front. Plant Sci.* **2024**, *15*, 1343928. [CrossRef] [PubMed]
35. Pan, T.; Chen, X.-L.; Hao, Y.-P.; Jiang, C.-W.; Wang, S.; Wang, J.-S.; Wei, Q.; Chen, S.-J.; Yu, X.-S.; Cheng, F.; et al. Optimization of Factors Affecting the Rooting of Pine Wilt Disease Resistant Masson Pine (*Pinus massoniana*) Stem Cuttings. *PLoS ONE* **2021**, *16*, e0251937. [CrossRef]
36. Enders, T.A.; Strader, L.C. Auxin Activity: Past, Present, and Future. *Am. J. Bot.* **2015**, *102*, 180–196. [CrossRef]
37. Pagnussat, G.C.; Lanteri, M.L.; Lombardo, M.C.; Lamattina, L. Nitric Oxide Mediates the Indole Acetic Acid Induction Activation of a Mitogen-Activated Protein Kinase Cascade Involved in Adventitious Root Development. *Plant Physiol.* **2004**, *135*, 279–286. [CrossRef]

38. Braun, L.; Wyse, D. Optimizing IBA Concentration and Stem and Segment Size for Rooting of Hybrid Hazelnuts from Hardwood Stem Cuttings. *J. Environ. Hortic.* **2019**, *37*, 1–8. [CrossRef]
39. García, M.; García, G.; Parola, R.; Maddela, N.R.; Pérez-Almeida, I.; Garcés-Fiallos, F.R. Root-Shoot Ratio and SOD Activity Are Associated with the Sensitivity of Common Bean Seedlings to NaCl Salinization. *Rhizosphere* **2024**, *29*, 100848. [CrossRef]
40. Rout, G.R. Effect of Auxins on Adventitious Root Development from Single Node Cuttings of *Camellia sinensis* (L.) Kuntze and Associated Biochemical Changes. *Plant Growth Regul.* **2006**, *48*, 111–117. [CrossRef]
41. Liao, C.Y. Advances on Internal Influence Factors of Adventitious Rooting of Forest Trees. *Agric. Sci. Technol.* **2017**, *18*, 1168–117[illegible]
42. Quan, J.; Ni, R.; Wang, Y.; Sun, J.; Ma, M.; Bi, H. Effects of Different Growth Regulators on the Rooting of Catalpa Bignonioide Softwood Cuttings. *Life* **2022**, *12*, 1231. [CrossRef] [PubMed]
43. Bai, T.; Dong, Z.; Zheng, X.; Song, S.; Jiao, J.; Wang, M.; Song, C. Auxin and Its Interaction with Ethylene Control Adventitious Root Formation and Development in Apple Rootstock. *Front. Plant Sci.* **2020**, *11*, 574881. [CrossRef]
44. Van Praet, S.; Rizza, A.; Tang, B.; Xie, C.; Jones, A.; Inzé, D.; Vanholme, B.; Depuydt, S. Coumarin Promotes Hypocotyl Elongation in Light-Grown *Arabidopsis thaliana* Seedlings by Enhancing Brassinosteroid Signalling in an Auxin-Dependent Manner. *J. Plant Growth Regul.* **2025**, *44*, 3290–3303. [CrossRef]
45. Welch, N.; Richter, C.; Franklyn-Miller, A.; Moran, K. Principal Component Analysis of the Biomechanical Factors Associated with Performance During Cutting. *J. Strength Cond. Res.* **2021**, *35*, 1715. [CrossRef]
46. Ning, X.; Su, J.; Zhang, X.; Wang, H.; Guan, Z.; Fang, W.; Chen, F.; Zhao, S.; Zhang, F. Evaluation of Volatile Compounds in Tea Chrysanthemum Cultivars and Elite Hybrids. *Sci. Hortic.* **2023**, *320*, 112218. [CrossRef]
47. Mirsoleimani, A.; Zinati, Z.; Abbasi, S. New Insights into the Identification of Biochemical Traits Linked to Rooting Percentage in Fig (*Ficus carica* L.) Cuttings. *J. Berry Res.* **2024**, *14*, 227–245. [CrossRef]
48. Singh, K.K. A Review: Multiplication of *Bougainvillea species* through Cutting. *ResearchGate* **2018**, *6*, 1961–1965.
49. Rao, L.M. Effect of Certain Plant Growth Regulating Substances on the Rooting of *Bougainvillea spectabilis* L. Stem Cuttings. *J. Jpn Soc. Hortic. Sci.* **1967**, *36*, 445–448. [CrossRef]
50. Salehi, M.; Zangeneh, M. Root Dynamics in *Bougainvillea glabra* Choisy Cuttings: Analyzing the Influence of Auxin Type, Concentration, and Cutting Type Over Time in Two Varieties. *J. Ornam. Plants* **2025**, *15*, 67–79. [CrossRef]
51. González-García, M.P.; Conesa, C.M.; Lozano-Enguita, A.; Baca-González, V.; Simancas, B.; Navarro-Neila, S.; Sánchez-Bermúdez, M.; Salas-González, I.; Caro, E.; Castrillo, G.; et al. Temperature Changes in the Root Ecosystem Affect Plant Functionality. *Plant Comm.* **2023**, *4*, 100514. [CrossRef] [PubMed]
52. Taduri, S.; Bheemanahalli, R.; Wijewardana, C.; Lone, A.A.; Meyers, S.L.; Shankle, M.; Gao, W.; Reddy, K.R. Sweetpotato Cultivars Responses to Interactive Effects of Warming, Drought, and Elevated Carbon Dioxide. *Front. Genet.* **2023**, *13*, 1080125. [CrossRef] [PubMed]

MDPI AG
Grosspeteranlage 5
4052 Basel
Switzerland
Tel.: +41 61 683 77 34

Horticulturae Editorial Office
E-mail: horticulturae@mdpi.com
www.mdpi.com/journal/horticulturae

www.ingramcontent.com/pod-product-compliance
Lightning Source LLC
Chambersburg PA
CBHW062048220926
51850CB00010B/2700